SpringerBriefs in Modern Perspectives on Disability Research

Series Editors

Gabriel Bennett, Independent Researcher
Klemzig, Australia

Emma Goodall, Healthy Possibilities
Seaford, Australia

This book series on disability research is a comprehensive collection of research on disability and related issues. The series is designed to promote interdisciplinary collaboration and exchange, bringing together scholars and practitioners from different fields to share their perspectives and insights. Disability research is an interdisciplinary field that examines the social, cultural, historical, and political dimensions of disability. It encompasses a wide range of topics, including disability rights, accessibility, assistive technologies, healthcare, education, employment, and social welfare. Disability research scholars employ a range of theoretical and methodological approaches to understand the experiences of people with disabilities, as well as the ways in which disability intersects with other social identities such as race, gender, sexuality, and class.

The series seeks to advance knowledge and understanding of disability by publishing rigorous, innovative, and relevant research. It aims to promote disability rights and social justice by highlighting the ways in which people with disabilities are marginalized and discriminated against in society, and advocating for greater social inclusion and accessibility. The series also seeks to inform policy and practice by disseminating research findings that can help to shape policy decisions and contribute to positive social change.

Anu Shibi Anilkumar •
Ramakrishnan Veerabathiran

Bioethics and Disability

Controversies in Genetic Engineering, Euthanasia, and Reproductive Rights

Anu Shibi Anilkumar
Human Cytogenetics and Genomics Laboratory, Faculty of Allied Health Sciences
Chettinad Hospital and Research Institute, Chettinad Academy of Research and Education
Kelambakkam, Tamil Nadu, India

Ramakrishnan Veerabathiran
Human Cytogenetics and Genomics Laboratory, Faculty of Allied Health Sciences
Chettinad Hospital and Research Institute, Chettinad Academy of Research and Education
Kelambakkam, Tamil Nadu, India

ISSN 3004-9709 ISSN 3004-9717 (electronic)
SpringerBriefs in Modern Perspectives on Disability Research
ISBN 978-981-95-8196-2 ISBN 978-981-95-8197-9 (eBook)
https://doi.org/10.1007/978-981-95-8197-9

This Springer imprint is published by the registered company Springer Nature Singapore Pte Ltd.
The registered company address is: 152 Beach Road, #21-01/04 Gateway East, Singapore 189721, Singapore

Preface

This book, *Bioethics and Disability: Controversies in Genetic Engineering, Euthanasia, and Reproductive Rights,* examines the rapidly evolving intersection between disability, ethics, and technological innovation in the twenty-first century. Advances in genetics, artificial intelligence, neurotechnology, reproductive medicine, and global health policy are reshaping how societies interpret human variation, autonomy, and justice. At the same time, people with disabilities continue to face entrenched inequities, discrimination, and systemic barriers that limit their access to healthcare, participation, and dignity. This work addresses the need to connect scientific progress with ethical responsibility and disability rights.

The book begins by situating disability within its social, historical, and political context, acknowledging the lasting impact of exclusion, institutionalization, eugenics, and paternalistic attitudes. It highlights the shift toward rights-based thinking, informed by international frameworks such as the UN Convention on the Rights of Persons with Disabilities, which emphasizes dignity, equality, and inclusion. It then examines key ethical challenges emerging from contemporary genetic and reproductive technologies. Developments such as CRISPR, preimplantation testing, and non-invasive prenatal screening provide significant medical possibilities, yet they also raise fundamental questions about normalization, choice, and the value placed on disabled lives. These concerns are closely tied to broader debates on enhancement, determinism, and the future of human diversity.

The discussion also considers global health systems, clinical practices, and disability laws, showing how policies and cultural attitudes shape access, fairness, and the quality of care across regions. Issues of accessibility, bias, and clinical injustice remain widespread, underscoring the need for disability-competent healthcare and inclusive governance. Emerging technologies, including AI, assistive innovations, and neurotechnologies, offer new forms of support and independence but introduce ethical tensions related to data governance, algorithmic discrimination, mental privacy, and equitable design. These developments reinforce the importance of ensuring that disabled communities are involved not only as beneficiaries but also as central contributors to research and policy.

Ultimately, this book argues that ethical progress requires more than technological advancement. It demands a commitment to justice, inclusive decision-making, and the recognition of lived experience as a form of expertise. By integrating scientific developments with disability-centered ethics, this work encourages readers to rethink assumptions and envision a future in which human diversity is fully respected and supported. This work aims to provide scholars, clinicians, policymakers, technologists, and advocates with an integrated understanding of disability within contemporary bioethical debates. It invites readers to reflect critically, to challenge inherited assumptions, and to imagine a future in which human diversity is respected not only in principle but in practice.

Chennai, Tamil Nadu
November 2025

Anu Shibi Anilkumar
Ramakrishnan Veerabathiran

Declarations

Funding: Not applicable

Availability of data and material: Not applicable

Code availability: Not applicable

Ethics approval: Not applicable

Consent to participate: Not applicable

Consent for publication: All authors have read and approved the manuscript.

Acknowledgment

The authors thank the Chettinad Academy of Research Education for their constant support and encouragement.

Competing Interests The authors have no competing interests to declare that are relevant to the content of this manuscript.

About the Book

Bioethics and Disability: Controversies in Genetic Engineering, Euthanasia, and Reproductive Rights delves into some of the most ethically challenging and socially significant debates of our time. It critically examines how emerging biotechnologies and healthcare policies affect the lives, rights, and representations of individuals with disabilities. Through the lens of bioethics, the book investigates controversial practices such as selective reproduction, prenatal screening, gene editing (including CRISPR), physician-assisted dying, and access to reproductive healthcare. It raises crucial questions about autonomy, personhood, quality of life, and the societal definition of "normalcy," arguing that disability perspectives are essential to shaping ethical frameworks in medicine and science.

This interdisciplinary volume brings together insights from disability studies, philosophy, medical ethics, law, and lived experiences to challenge ableist assumptions often embedded in bioethical discourse. It offers a nuanced, inclusive, and justice-oriented approach to ethical decision-making, advocating for policies and practices that respect human diversity and dignity. Suitable for academics, students, bioethicists, healthcare professionals, and policymakers, this book encourages more profound reflection on the implications of scientific advancement and promotes dialogue on how to ethically navigate the future of healthcare and human rights, in ways that include rather than marginalize people with disabilities.

Contents

Abbreviations

ACA	Affordable Care Act
ADA	Americans with Disabilities Act
AI	Artificial intelligence
ART	Assisted reproductive technology
AT	Assistive technology
BCI	Brain–computer interface
CI	Cochlear implant
CRISPR-Cas9	Clustered regularly interspaced short palindromic repeats
CRPD	Convention on the Rights of Persons with Disabilities
DNR	Do not resuscitate
DXA	Dual-energy X-ray absorptiometry
ECG/EEG/EMG	Cardiac/brain/muscle electrical activity tests
EHR	Electronic health records
FES	Functional electrical stimulation
GDPR	General Data Protection Regulation
HLA	Human leukocyte antigen
ICSI	Intracytoplasmic sperm injection
ICT	Information and communication technology
ICU	Intensive care unit
IVF	In vitro fertilization
LMIC	Low- and middle-income countries
MRI/CT	Magnetic resonance imaging/computed tomography
NCD	National Council on Disability
NIEPID	National Institute for the Empowerment of Persons with Intellectual Disabilities
NIPT	Noninvasive prenatal testing
NTA	National Trust Act (India)
PwD/PWD	Persons with disabilities
UNCRPD	United Nations Convention on the Rights of Persons with Disabilities

UNESCO	United Nations Educational, Scientific and Cultural Organization
UNICEF	United Nations International Children's Emergency Fund
PGD	Preimplantation genetic diagnosis
PGS	Preimplantation genetic screening
PGT/PGT-A/PGT-M	Preimplantation genetic testing (aneuploidy/monogenic)
PGx	Pharmacogenomics
QALY	Quality-adjusted life year
RPwD Act	Rights of Persons with Disabilities Act (India)
SOGIESC	Sexual Orientation, Gender Identity, Expression & Sex Characteristics
UDBHR	Universal Declaration on Bioethics and Human Rights
WHO	World Health Organization
WIOA	Workforce Innovation and Opportunity Act

Chapter 1
Foundations: Bioethics, Disability, and Historical Injustices

Abstract Disability refers to any physiological or cognitive condition that limits an individual's ability to perform regular duties or interact with others socially. This chapter explores how the perception of disability has evolved from one of marginalization and stigma to a framework centered on rights, integrity, and moral inclusion. It examines the evolution of disability models, encompassing medical, sociological, and bioethical perspectives, and their influence on attitudes, laws, and research methodologies. The impact of historical injustices on contemporary disability ethics, including forced sterilization, institutionalization, and Nazi eugenics initiatives, is examined. Intersectionality serves as a framework that connects disability with other marginalized identities to combat systemic injustices. The UN Convention on the Rights of Persons with Disabilities (CRPD) acts as a standard for international justice and self-determination. Altogether, this chapter highlights the inevitability of intersectional, ethically upright, and rights-based strategies to ensure the representation, equity, and dignity of people with disabilities.

Keywords Disability Models · Bioethics · Intersectionality · Forced sterilization · Eugenics · Reparative justice

1.1 Introduction

Any physical or mental limitation that prevents a person from participating in everyday activities and social interactions in their surroundings is considered a disability (Babik & Gardner, 2021). An estimated 16% of the world's population, or 1.3 billion people, are living with a significant impairment (WHO, 2022). The United Nations International Children's Emergency Fund (UNICEF) reports that millions of children worldwide live with moderate-to-severe disabilities (Olusanya et al., 2022). The United Nations Convention on the Rights of Persons with Disabilities (UNCRPD), the first universal human rights convention ratified in the twenty-first

A. Shibi Anilkumar, R. Veerabathiran, *Bioethics and Disability*, SpringerBriefs in Modern Perspectives on Disability Research,
https://doi.org/10.1007/978-981-95-8197-9_1

century, established a framework of equality and dignity for persons with disabilities (PwD) in international law (Ashalatha et al., 2024).

Disability was often associated with stigma, dread, or superstition in ancient societies. Despite continued marginalization, religious care and philanthropic organizations for PwD began to emerge during the Middle Ages. More compassionate views of disability were made possible by the Enlightenment, which later introduced concepts of reason, human value, and social duty. Historical periods collectively influenced the gradual shift in how people view disability (Kumar & Lal, 2025).

People with disabilities (PwD) have been excluded and abused on a large scale. The ancient Greeks abandoned disabled newborns, used for amusement by the Romans, and even killed during the Renaissance. In colonial America, they were either caged for public display, burned, or drowned. In nineteenth-century American asylums and European orphanages, many people were shackled or kept in isolation due to mistreatment and dehumanization. The twentieth century saw Nazi crimes, eugenics, and forced sterilization. People with disabilities (PwD) faced stigma, abuse, institutionalization, and educational denials worldwide, exposing a long history of injustice in all civilizations (Barton & Oliver, 1997).

Throughout Western history, disability has arisen at the interface of the specific needs of an impairment and the broader political and economic framework surrounding disability (Braddock & Parish, 2001). Disability history emerged as a distinct field in the 1980s, mainly due to the UN's 1981 International Year of Disabled Persons and the Disability Rights Movement. As researchers unearthed the hidden histories of PwD to dispel prejudices and emphasize their agency, an activist-driven perspective that still defines disability history today, the discipline grew in the 1990s, particularly following the passage of the Americans with Disabilities Act (ADA) in 1990 (Blackie & Moncrieff, 2022).

A systematic approach to recognizing, evaluating, and resolving problems in healthcare and the biomedical sciences that prioritizes PwD first-hand accounts is known as disability bioethics (Guidry-Grimes 2022). The origins and development of bioethics are closely tied to the history of disability (Mukherjee et al., 2022). Although bioethics took a while to address disability, it doesn't stop bioethicists from actively influencing the subject. In addition to critically analyzing the rights-based framework that has largely governed disability activism, bioethics now plays a significant role in advancing more ethical and inclusive treatment and research practices (Kuczewski, 2001).

When considering some of the most controversial topics surrounding human life, such as selective abortion, euthanasia, the possible eugenic effects of prenatal testing, and genetic technological advancements, it becomes even more necessary and complicated to discuss ethics from a disability viewpoint (Jarman, 2008). Due to the Human Genome Project and its associated program, "Ethical, Legal, and Social Issues" (ELSI), the field of bioethics has undergone significant expansion since the 1970s. Many bioethicists have discussed the nature of disability and the lives of individuals with disabilities to support specific suggestions for genetic policy as a result of the developing genetic technologies (Amundson & Tresky, 2008). This

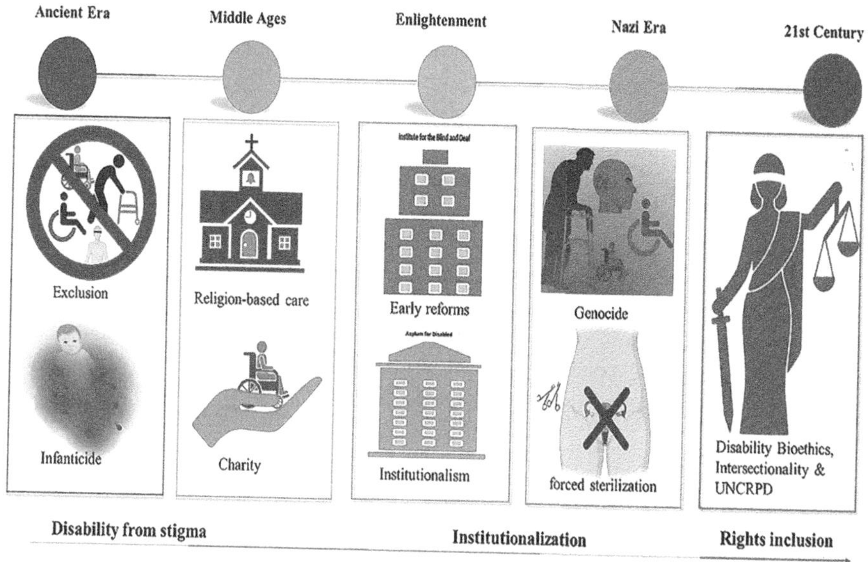

Fig. 1.1 Historical Progression of Disability Perceptions and Treatment. This figure illustrates the historical evolution of disability treatment and societal attitudes toward persons with disabilities, progressing from exclusion to rights-based inclusion. The first section depicts early discrimination, neglect, and infanticide of disabled infants, symbolized by a prohibition sign over mobility aids and a distressed infant. The second section represents the emergence of religious charity and basic care institutions, as evidenced in church-based support and protective caregiving imagery. The third section highlights the rise of specialized schools, blind and deaf institutes, and asylums that, although providing structured environments, often reinforced segregation. The fourth section portrays twentieth-century injustices, including forced sterilization and deeply embedded societal stigma against disability. The final section signifies modern disability rights and bioethics, represented by the figure of Lady Justice holding scales, symbolizing equality, legal protections, and ethical responsibility. Overall, the image narrates the shift from historical marginalization to contemporary frameworks grounded in dignity, justice, and disability rights

chapter examines the revolution in the perception of disability, shifting from a stigmatized ailment to a rights-based perspective, and highlighting the impact of social power and injustice on its evolving global inclusion. The historical progression of perceptions and treatment of disability from the ancient era to the twenty-first century is shown in Fig. 1.1.

1.2 Disability Models

Disability models serve as records that provide information on how people with and without disabilities interact in society, and also guide our responses to variations in the body and brain, which have important implications for both groups. The way that individuals with disabilities are treated legally, their access to education and work opportunities, and how they are portrayed in the media and public consciousness are all significantly impacted by models of disability (Zaks, 2024).

Numerous conceptual models or frameworks characterize disability, including the medical model, the charity model, the social model, and others. These models and frameworks differ significantly in their dimensions and levels of aspects and components. Since there is no universally accepted definition of impairment, researchers will inevitably turn to a specific model of disability that is more suited to the situation they are studying (Retief & Letšosa, 2018). The development of these models is mainly based on scientific understanding (Pezdek & Rasiński, 2017). It has not always been simple to reconcile ethics, especially bioethics, with disability. The underlying premise of mainstream bioethics appeared to be that disabilities cruelly reduce a person's quality of life, making disability almost a guarantee of a life with less significance (Bickenbach, 2012).

1.2.1 Medical Model

The medical model, a traditional model of disability, views disability as an individual impairment requiring medical treatment. It promotes a disease-oriented approach to the patient by emphasizing physiological, biochemical, and anatomical dysfunction (Jeffery, 2006). The model has traditionally supported negative attitudes and stereotypical perceptions of PwD by framing disability as an individual's deficit instead of a societal issue (Sofokleous & Stylianou, 2023).

The medical model associates disability with dysfunction, disregarding the influence of contextual and environmental factors (Davies & Soni, 2025). Medical professionals often treat individuals with impairments from this perception. This approach highlights personal limitations, does not encourage patient-centeredness, and may hinder medical professionals' capability to understand or accommodate the social and ecological aspects of disability (Phillips et al., 2021). Maintaining the status quo in society by keeping established power structures is one of the central resolutions of the medical model. Subsequently, the medical model does not challenge prevailing systems of discrimination and social expectations; instead, it reinforces them, depoliticizing disability (Beaudry, 2025).

1.2.2 Social Model

In the 1970s, disability activism gave rise to the social model of disability, which was further refined by academics such as Michael Oliver and Vic Finkelstein. It opposed the medical model, which viewed disability as a personal flaw that needed to be fixed by medicine. The social model claimed that discrimination and social impediments, rather than a person's physical or mental deficiency, are the root cause of disability (Beaudry, 2016).

In the latter half of the twentieth century, many disabled individuals politicized disability due to a variety of circumstances, including poverty, isolation, medical dominance and intervention, institutionalization, incarceration, and mass debilitating events like World War II. The Union of the Physically Impaired Against Segregation (UPIAS), founded in 1974 by disability activists disillusioned with the status quo, is regarded as one of the most influential organizations in the history of the social model of disability. The social model was rapidly adopted across activist and policy contexts (Thorneycroft, 2024).

According to a disability rights approach in bioethics, disability should be seen as a social justice issue that necessitates societal change rather than just a medical or technological problem requiring clinical remedies. End-of-life decisions, healthcare resource allocation, genetic testing and therapy, research involving PwD, treatment refusal, selective care of newborns, and discussions on personhood and disability-adjusted life years are just a few of the primary bioethical debates that directly affect PwD. The fundamental bioethical concepts of autonomy, beneficence, nonmaleficence, and justice must be reexamined in these discussions using a framework that takes into account the rights and real-life experiences of individuals with disabilities (Wolbring, 2003).

1.2.3 Bioethical Model

Ethics is made more challenging by the complexity of bioethical difficulties and the propensity of one ethical question to lead to others (Savulescu, 2024). The four guiding principles of ethics are autonomy, justice, nonmaleficence, and beneficence (Varkey, 2021). Over the past few decades, there have been numerous conflicts between the disability movement and the discipline of bioethics (Anilkumar & Veerabathiran, 2025), especially when it comes to life-or-death matters like assisted suicide for disabled persons or prenatal selection. Despite these conflicts, a disability viewpoint has been openly included in the work of some feminist ethicists, and vice versa. Approaches to bioethics that take into account the theoretical and empirical insights provided by PwD are known as the Disability Bioethics Model. To improve the situation for PwD, these methods diligently incorporate those findings into their theories, methodologies, and inquiries (Vanaken, 2022).

The goal of the bioethical principle is to guarantee equitable access to opportunities, services, and participation in society for those with disabilities. Bioethicists guarantee that population health purposes and individual rights are stable while improving a more inclusive view of what it means to lead a satisfying life with a disability. Exercising reproductive decisions is an additional part where bioethics and disability overlap. The ethical consequences of screening for and possibly finding specific disabilities raise questions of prejudice, eugenics, and the importance of diversity in the human race. It is thus significant to concentrate on the perspectives and experiences of individuals with disabilities, recognizing the variability within the disability group, and work toward laws and procedures that maintain

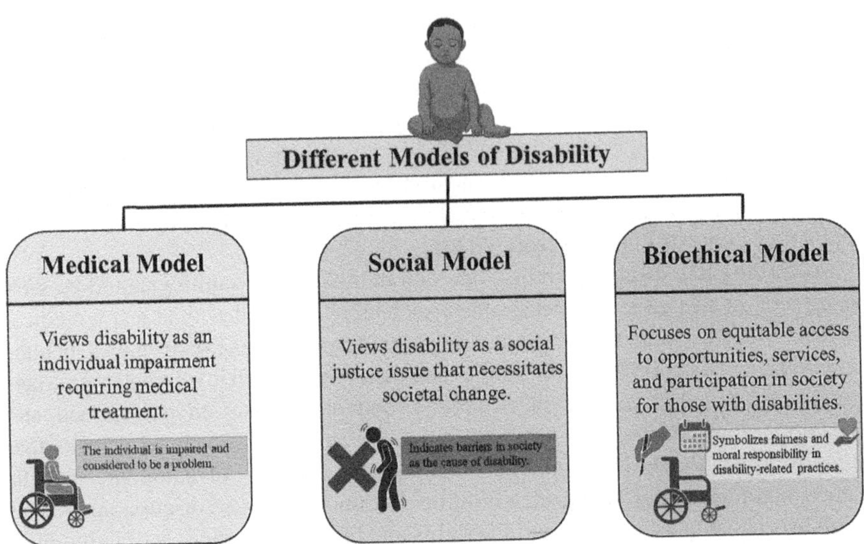

Fig. 1.2 Overview of the medical, social, and bioethical models of disability. The figure presents three models of disability: Medical, Social, and Bioethical, displayed as three colored panels beneath a banner titled "Different Models of Disability," with an illustration of a child above the banner. The Medical Model panel states that disability is viewed as an individual impairment requiring medical treatment, and it includes an icon of a wheelchair user with a label indicating that the individual is considered the problem. The Social Model panel describes disability as a social justice issue caused by societal barriers, shown with an icon of a person facing an obstruction. The Bioethical Model panel emphasizes equitable access, moral responsibility, and fairness in disability-related practices, accompanied by icons symbolizing support and justice

impartiality, self-respect, and fairness for all while navigating the connection between bioethics and disability (Ikecukwu, 2024). Figure 1.2 compares three dominant perspectives on disability, highlighting how each model interprets the causes, responsibilities, and societal implications of disability.

1.3 Intersectionality and Ethical Frameworks

Intersectionality is a significant context that allows the exploration of social injustices that continue across physical and disciplinary domains through studies and practice (Ovalle et al., 2023). It serves as an empirical and systematic tool, a method, and a disposition, with roots in Black feminism and Critical Race Theory. The term was coined in 1989 by Kimberlé Crenshaw to highlight the marginalization of Black women within feminist and anti-racist movements, as well as in anti-discrimination law (Carbado et al., 2013). Crenshaw's concept has since been extended to comprise several dimensions of transformation and exclusion (Watermeyer & Swartz, 2023).

Although it originated from feminist and anti-racist studies, intersectionality is now applied in a diverse range of fields and contexts. Its fundamental concepts have been extensively accepted, leading to diverse methodological approaches across different areas of study (Brünig et al., 2024). As intersectionality gained importance, feminist researchers and activists began urging governments and international organizations to identify and address interrelated forms of oppression and discrimination. Accordingly, there is growing recognition of intersectional or multi-ground discrimination by international human rights bodies.

Despite this progress, studies on intersectionality in global human rights law, particularly those related to disability, remain limited. Since the implementation of the Convention on the Rights of Persons with Disabilities (CRPD) in 2006, several researchers have investigated the intersection of disability with other forms of discrimination (De Beco, 2020). However, even with these developments, the CRPD does not fully account for the variability among PwD. Many groups, specifically disabled members of racial or ethnic minorities, even depend on other human rights treaties to state their rights, as intersectional aspects such as race and disability often receive inadequate attention (De Beco, 2020; Ruiz, 2025).

Most disability studies are rooted in psychology's traditional medical model, which also views disability primarily as a pathology rather than a social construct. Applying intersectionality enables a shift toward examining the external social forces that shape perception. A prominent instance is the 2001 Ontario Human Rights Commission report, which projected a framework for addressing numerous discrimination claims and emphasized that legal systems frequently fail to address the sufferers of multiply marginalized individuals with disabilities (Shaw et al., 2012).

According to Barned et al. (2019), intersectionality in bioethics necessitates that bioethicists reflect on their individual social positions and political assumptions, as these influence their ethical reasoning. To foster righteousness and equality, they emphasize the need to transform existing practices and institutional procedures. Utilitarianism, which advocates for the greatest happiness for the greatest number, has had a profound influence on public policy. However, this rationale can make specific communities powerless or fragile, including unborn individuals, those with impairments, and people who are approaching death. This problem raises ethical conflicts regarding society's duty toward people across the lifespan, particularly those with impairments (Hilliard, 2011).

Utilitarian ethics, being an outcome-based framework, assesses the ethical nature of actions based on their results, appreciating everyone's happiness impartially and reinforcing common welfare. Hence, individual ambitions should be pursued only if doing so ensures total benefit. In contrast, deontic ethics holds that some actions are inherently right or wrong, independent of their consequences, a perspective that justifies our commitment to human rights (Playford et al., 2015).

The ethical basis of clinical practice for individuals with severe impairments should be grounded in the principles of beneficence, nonmaleficence, autonomy, and justice. Ethical reasoning within medical ethics is directed by three key philosophical approaches: deontology, consequentialism, and virtue ethics. In overall aspects, Deontology stresses adherence to rules, laws, and institutional norms; consequentialism (or utilitarianism) focuses on the balance of benefits and harms; while virtue ethics, deeply rooted in Aristotle's philosophy, highlights justice and moral discernment. Within these, virtue ethics is constantly considered the most suitable framework for healthcare services for PwD (Rachmawati et al., 2023).

1.4 United Nations Convention on the Rights of Persons with Disabilities (UNCRPD) and Global Ethical Obligations

United Nations Convention on the Rights of Persons with Disabilities (UNCRPD), which opened for ratification in 2007 (Veerabathiran & Thomas, 2025), establishes minimum standards for the rights of individuals with disabilities in several domains. More than 185 states have endorsed the Convention to date (Grech et al., 2023). Protection and advancement of the civil, political, social, economic, and cultural rights of individuals with disabilities are outlined in the UNCRPD (Shikako et al., 2023). The guiding concepts include equality of opportunity, accessibility, nondiscrimination, respect for one's inherent dignity, individual autonomy (including the ability to make one's own decisions), and complete and adequate engagement and inclusion in society. The UNCRPD's core principle is the incorporation of PwD into communities and society. The UNCRPD comprises 50 articles that expand on these eight principles in relation to other aspects of society, including employment, education, and health (Steinert et al., 2016).

The concepts of the social model of disability work as the intellectual foundation for the CRPD. The UN has created six human rights principles to direct rights-based strategies. These comprise inseparability, inclusiveness, inalienability, mutual reliance and interrelatedness, equal opportunity and equality, inclusion and involvement, and responsibility and justice (Lang et al., 2011). In line with Article 24 of the Convention, states that have accepted and signed the CRPD are entitled to offer more resources and energy toward the education of children with disabilities (Wescott et al., 2023). Article 3 specifies eight principles that member states parties must comply with to transform the way PwD are treated (McCusker et al., 2023). To overcome institutional constraints that limit fair access to human rights, both the CRPD and social service practitioners share a common understanding of the social model of disability. This approach is evident in CRPD Article 12, which asserts that PwD must be protected by active safeguards that recognize their rights, will, and preferences (12.4), as well as have the right to lawful recognition (12.1), legal competence on an equal basis with others (12.2), and assistance to the support required to exercise that capacity (12.3) (McCusker et al., 2023).

With their emphasis on protecting against misuse and preserving autonomy, the Nuremberg Code (1947) and the Declaration of Helsinki (1964), with subsequent modifications) have shaped the existing framework of human research ethics. These values are consistent with the CRPD, which supports noncoercive practices (Article 4), autonomy (Article 3), and freedom of choice. Justice, equality, and human rights are now central to recent ethical discussions, and the CRPD and disability health studies are intimately related. The International Classification of Functioning and the CRPD both recognize that everyone is vulnerable and interdependent, yet existing ethical governance frameworks, despite their protective nature, are frequently criticized as paternalistic and inconsistent with inclusive research principles (Durham et al., 2014).

1.5 The Legacy of Forced Sterilization and Institutionalization

Forced sterilization is one of the most pervasive human rights violations in the world today, having been legalized in many parts of the world since the second half of the nineteenth century. Sterilization has been deemed forced, coerced, or involuntary if it is carried out without the individual's full, free, and informed consent, despite an express refusal, without the individual's knowledge of the intervention, or without the chance for the individual to offer consent (Yupanqui-Concha et al., 2021). "A process or act that renders an individual incapable of sexual reproduction" is the definition of sterilization. When a person is sterilized after expressly declining the procedure, when it occurs without her awareness, or when she is not given the chance to give consent, it is deemed to be "forced." Forced sterilization, carried out under the name of valid medical treatment or with the approval of others, is especially dangerous for women with disabilities (Serrato Calero et al., 2021).

Lawmakers in many nations were persuaded by eugenics arguments in the late 19th and early twentieth centuries that unrestrained reproduction of "the feeble-minded" and other "undesirables" posed a significant threat to civilization. Sex-segregated institutions and, in many cases, compulsory sterilization were enshrined in the eugenic schema. These practices of eugenics are not as widely accepted. But the practice of eugenics did not end (Raposo, 2022). The eugenics movement aimed to isolate and eradicate individuals with mental and intellectual deficiencies, in addition to those with obvious physical impairments (McDaniel & Friedman, 2021).

People with disabilities (PwD) were frequently kept apart at home rather than being integrated into communities in the United States, while in the nineteenth century, they were dehumanized in asylums and orphanages in Europe. The use of restraints, often due to staff shortages, and institutions, such as Massachusetts' "Institution for Idiots," were examples of pervasive negligence. The twentieth century brought with it worldwide institutionalization, exclusion from schools, and cruel "treatments," as well as forced sterilization, eugenics, and Nazi crimes. The ongoing need for encouragement and social revolution in the way disabled people are treated is stressed by these inequalities (Dineshan & Geetha, 2024).

The majority of research has fixated on regulations and eugenics discussions, paying limited regard to the experiences of the impacted women. As reported by some research, when women with intellectual disabilities discover they were sterilized against their will, they frequently experience dejection, rage, and desolation. Their ability to provide their approval was previously ignored. Nowadays, supported decision-making is supported, and capability is presumed by the Mental Competence Act 2005 (England & Wales). But even if medical practices have changed, concerns involving the autonomy and reproductive rights of women with intellectual disabilities still subsist (Tilley et al., 2012).

The deinstitutionalization guideline is outlined in Article 19 of the UN Convention on the Rights of Persons with Disabilities, which emphasizes the transition from institutional care systems to person-centered services. This notion remains imperative to modern disability legislation and implementation. However, "deinstitutionalization" has evolved over time and, in many circumstances, its current global meaning is nuanced and frequently debated. Historically, drastic tactics like sterilization were used to restrict fertility, sexuality, and potential motherhood, and institutional segregation was also implemented along gender lines to prevent the reproduction of people deemed to be "moral defectives" (Roets et al., 2022).

1.6 Nazi Eugenics Programs

The current form of eugenics philosophy was advanced by Francis Galton (1822–1911), who coined the name "eugenics" in 1883, although its roots can be found in Plato's Republic (c. 378 BCE), which promotes selective breeding to advance the human race. Galton and others believed that eugenics was a scientifically supported governmental initiative. Proponents of eugenics initially separated positive and negative eugenics. The goal of positive eugenics is to encourage the reproduction of "superior individuals" from "good stock." Refusing to procreate offspring of "defective stock" is known as negative eugenics (Jason, 2025).

Sex education, marriage counseling, marital limitations, contraceptive programs, sterilization, abortion, euthanasia, and finally murder were all examples of eugenic methods (Da Silva & Hubbard, 2024). The eugenics movement used harsh terms to characterize those thought to possess undesirable traits, including the mentally sick, the impoverished, sexually promiscuous, alcoholics, epileptics, sexual deviants or delinquents, criminals, mentally ill persons, imbeciles, n'er do well, and anyone who wasn't White (Charlot et al., 2022).

Hitler was famously enthusiastic about eugenics. He read the standard Weimar-era eugenics textbook Menschliche Erblichkeitslehre und Rassenhygiene (Principles of Human Heredity and Racial Hygiene) and incorporated its concepts into Mein Kampf (My Struggle) (Grodin et al., 2018). During the Nazi regime, euthanasia was carried out in various forms and at different stages. One of the most notorious programs, known as Aktion T4, named after Tiergartenstrasse 4, was a large-scale, systematic, and secret operation aimed at the extermination of people with disabilities. Under this program alone, more than 70,000 psychiatric patients from both state-run and private hospitals and asylums were transported to killing centers and murdered.

The Nazis validated euthanasia as a way to "purify" the *Volk* by eradicating individuals deemed ineligible, prioritizing the benefit of the community over individual rights. This procedure of deindividualization gave rise to the objectification of patients, permitting Nazi T4 doctors to execute life-and-death decisions based on presumed economic output. Differing from the former forced sterilization campaign, Aktion T4 was designed to be a top-secret operation. Each killing centers functioned following strict confidentiality, maintaining its own internal law enforcement agency and domestic records to cover the brutal acts (Remington, 2023). A state-run policy of forced sterilization and murder would have been far more challenging to carry out without the voluntary cooperation of doctors, but collaboration between the Nazis and medical professionals provided a compelling rationale. Genocide was the ultimate result of what started as purification (Grodin et al., 2018).

Following World War II, several eugenic societies in various nations changed their names and modified their operations to disassociate themselves from the atrocities perpetrated. Eugenics was publicly denounced when these techniques were exposed. Consequently, there is currently a reasonable hesitancy to recognize the potentially eugenic elements of reproductive genetics (Dive & Newson, 2022). Nondiscrimination and the fundamental values of "diversity, equity, and excellence in clinical practice, research, education, and administration" are outlined in the American Association of Marriage and Family Therapy's (AAMFT) code of ethics (AAMFT, 2015). Eugenics became the scientific justification for several social ills and is incompatible with the AAMFT code of ethics' declared core objective of "Diversity, Equity, and Excellence" (AAMFT, 2015, p. 113).

Eugenic thought has ranged from rejecting state involvement in reproduction to advocating limited state roles through information or subsidies, and even strong state control through compulsion. Modern bioethicists unanimously reject Nazi-style eugenics, including sterilization and mass murder, although public perception still links eugenics with its worst abuses. Some eugenic principles persist, such as bans on incest and cousin marriage to prevent genetic disorders, and genetic counseling for at-risk groups to avoid diseases like Tay-Sachs, sickle-cell anemia, and thalassemia (Veit et al., 2021). These historical memories remain closely linked to contemporary bioethical debates and the evolution of human genetics.

1.7 Ethical Lessons and Reparative Justice

Although disability history is a relatively young field, it has grown notably in scope. Initially dominated by American historians focusing on the nineteenth and twentieth centuries, recent years have seen an increasing number of studies examining disability before 1800 (Blackie & Moncrieff, 2022). An emerging strategy to address historical injustices caused by archives that have ignored or misrepresented marginalized communities is known as reparative description. This movement has led to an increasingly accurate representation of disability history, with many archivists publishing reparative projects or theoretical frameworks that focus primarily on the result (Weiss, 2024).

Research on disabilities is influenced by constant debates about the various ways to interact with disabled individuals, all affected by cultural and historical contexts. Therefore, even within institutional ethics frameworks, there is no single "right" method. Researchers, as an alternative, unite field knowledge and context to reinterpret what it means to conduct ethical research (Jones, 2021). People with disabilities (PwD) are growingly participating in education, employment, social, and cultural events. Conventional approaches often focus on retrofitting physical spaces or offering limited accommodations. People with disabilities (PwD) are presently permitted to interact with the world on their own, overcoming restrictions that once constrained their entire contribution, by making use of AI capabilities (Almufareh et al., 2024). Restorative justice reinforces prevention and future advancement, serving as a means of achieving fair and equitable resolutions. Regarding responsibility, restorative justice can also help affected individuals, perpetrators, and societies (Saefudin et al., 2022).

In recent decades, diverse frameworks have redefined disability as a strength, rather than an insufficiency, cultivating independence from impairment. In contrast, many models also neglected to incorporate other aspects of identity (such as color, ethnicity, or sex) or address systemic and environmental factors like inclusivity. More comprehensive strategies, such as intersectional frameworks, provide meaningful insights in confronting health disparities. Although there is growing support for intersectionality in psychology, the connection of disability with other oppressed groups remains largely disregarded (Brinkman et al., 2023).

The advancement of gender studies is prominently shaped by intersectionality. The integration of research and social activism within academic communities has reshaped the field, extending its attention beyond sex and gender to incorporate perspectives from gender studies, postcolonial theories, and disability studies (Brünig et al., 2024). The CRPD framework finds gaps and constraints in laws and policies that affect the health of PwD. It also provides direction for ensuring that PwD can fully understand their rights (Ashalatha et al., 2024).

1.8 Conclusion

The history of disability shows a long and far journey from marginalization and oppression to CRPD recognition and empowerment. In modern research, medical services, and administration, it is essential to avoid ableism through acknowledging historical injustices such Nazi eugenics and forced sterilization. Restorative justice and reparative justice represent two initiatives that emphasize persistent attempts to achieve and promote responsibility. Understanding how disability interacts with broader social, cultural, and structural inequities is strengthened by the intersectional and bioethical frameworks. Advances like AI-driven inclusion demonstrate that technology can help enhance involvement and autonomy. Despite these progressions, persistent obstacles serve as a reminder that inclusive practice and ethical thought remain vital factors in disability research and governance. Maintaining equality, dignity, and justice for all PwD necessitates ongoing study of human rights, ethics, and history.

References

Almufareh, M. F., Kausar, S., Humayun, M., & Tehsin, S. (2024). A conceptual model for inclusive technology: Advancing disability inclusion through artificial intelligence. *Journal of Disability Research, 3*(1), 20230060. https://doi.org/10.57197/JDR-2023-0060

American Association of Marriage and Family Therapy's (AAMFT) code of ethics (AAMFT, 2015). https://healthy.arkansas.gov/wp-content/uploads/2015-AAMFT-code-of-ethics.pdf

Amundson, R., & Tresky, S. (2008). Bioethics and disability rights: Conflicting values and perspectives. *Journal of Bioethical Inquiry, 5*(2), 111–123. https://doi.org/10.1007/s11673-008-9096-3

Anilkumar, A. S., & Veerabathiran, R. (2025). Genomics and congenital disabilities. In *The Palgrave encyclopedia of disability* (pp. 1–13). Springer Nature. https://doi.org/10.1007/978-3-031-40858-8_510-1

Ashalatha, S. L., Kumar, S., Santosh, B. R., & Ravindra, B. K. (2024). Navigating UNCRPD concluding observations on Article 27: Policy exploration on disability inclusive employment. *Cogent Social Sciences, 10*(1), 2425169. https://doi.org/10.1080/23311886.2024.2425169

Babik, I., & Gardner, E. S. (2021). Factors affecting the perception of disability: A developmental perspective. *Frontiers in Psychology, 12*, 702166. https://doi.org/10.3389/fpsyg.2021.702166

Barned, C., Lajoie, C., & Racine, E. (2019). Addressing the practical implications of intersectionality in clinical medicine: Ethical, embodied and institutional dimensions. *The American Journal of Bioethics, 19*(2), 27–29. https://doi.org/10.1080/15265161.2018.1557278

Barton, L., & Oliver, M. (1997). *Disability studies: Past, present and future*. The Disability Press.

Beaudry, J. S. (2016, February). Beyond (models of) disability?. In *The Journal of Medicine and Philosophy: A Forum for Bioethics and Philosophy of Medicine* (Vol. 41, No. 2, pp. 210–228). Journal of Medicine and Philosophy Inc. doi:https://doi.org/10.1093/jmp/jhv063.

Beaudry, J. S. (2025). The medical model of disability. *The Oxford handbook of philosophy of medicine, 371*. doi:https://doi.org/10.1093/oxfordhb/9780197625835.013.0017.

Bickenbach, J. (2012). Ethics, disability and the international classification of functioning, disability and health. *American Journal of Physical Medicine & Rehabilitation, 91*(13), S163–S167. https://doi.org/10.1097/phm.0b013e31823d5487

Blackie, D., & Moncrieff, A. (2022). State of the field: Disability history. *History, 107*(377), 789–811. https://doi.org/10.1111/1468-229X.13315

Braddock, D. L., & Parish, S. L. (2001). History of disability. *Handbook of disability studies, 11*.

Brinkman, A. H., Rea-Sandin, G., Lund, E. M., Fitzpatrick, O. M., Gusman, M. S., Boness, C. L., & Scholars for Elevating Equity and Diversity (SEED). (2023). Shifting the discourse on disability: Moving to an inclusive, intersectional focus. *The American Journal of Orthopsychiatry, 93*(1), 50–62. https://doi.org/10.1037/ort0000653

Brünig, L., Kahrass, H., & Salloch, S. (2024). The concept of intersectionality in bioethics: A systematic review. *BMC Medical Ethics, 25*(1), 64. https://doi.org/10.1186/s12910-024-01057-5

Carbado, D. W., Crenshaw, K. W., Mays, V. M., & Tomlinson, B. (2013). INTERSECTIONALITY: Mapping the movements of a theory1. *Du Bois Review: Social Science Research on Race, 10*(2), 303–312. https://doi.org/10.1017/S1742058X13000349

Charlot, L. A., Washington, K., & Hall, C. (2022). Exhumed: Reckoning with the history of eugenics in marriage and family therapy. *The Family Journal, 30*(4), 493–498.

Davies, N., & Soni, A. (2025). Reflecting on the models of disability utilised in schools. *Support for Learning, 40*(3), 233–241. https://doi.org/10.1111/1467-9604.70002

Da Silva, S. M., & Hubbard, K. (2024). Confronting the Legacy of Eugenics and Ableism: Towards Anti-Ableist Bioscience Education. *CBE life sciences education, 23*(3), es7. https://doi.org/10.1187/cbe.23-10-0195

De Beco, G. (2020). Intersectionality and disability in international human rights law. *The International Journal of Human Rights, 24*(5), 593–614. https://doi.org/10.1080/13642987.2019.1661241

Dineshan, A., & Geetha, B. (2024). Concept of 'Disability' – From the historical trajectory. In *Proceedings of international conference on social sciences, languages and culture* (pp. 173–180).

Dive, L., & Newson, A. J. (2022). Reproductive carrier screening: Responding to the eugenics critique. *Journal of Medical Ethics, 48*(12), 1060–1067.

Durham, J., Brolan, C. E., & Mukandi, B. (2014). The Convention on the Rights of Persons with Disabilities: A foundation for ethical disability and health research in developing countries. *American Journal of Public Health, 104*(11), 2037–2043. https://doi.org/10.2105/AJPH.2014.302006

Grech, S., Weber, J., & Rule, S. (2023). Intersecting disability and poverty in the Global South: Barriers to the localization of the UNCRPD. *Social Inclusion, 11*(4), 326–337. https://doi.org/10.17645/si.v11i4.7246

Grodin, M. A., Miller, E. L., & Kelly, J. I. (2018). The Nazi physicians as leaders in eugenics and "Euthanasia": Lessons for today. *American Journal of Public Health, 108*(1), 53–57. https://doi.org/10.2105/AJPH.2017.304120

Guidry-Grimes, L. (2022). Disability bioethics and the commitment to equality. *Theoretical Medicine and Bioethics, 43*(4), 209–220. https://doi.org/10.1007/s11017-022-09575-2

Hilliard, M. T. (2011). Utilitarianism impacting care of those with disabilities and those at life's end. *The Linacre Quarterly, 78*(1), 059–071. https://doi.org/10.1179/002436311803888474

Ikecukwu, O. M. (2024). The intersection of bioethics and disability right. *Mexican Bioethics Review ICSA, 6*(11), 12–16. https://doi.org/10.29057/mbr.v6i11.13050

Jarman, M. (2008). Disability studies ethics: Theoretical approaches for the undergraduate classroom. *Review of Disability Studies: An International Journal, 4*(4) http://hdl.handle.net/10125/58368

Jason, G. J. (2025). Selling eugenic killing: The Nazi propaganda campaign to support the liquidation of the disabled.

Jeffery, A. (2006). Moving away from the medical model. *Veterinary Nursing Journal, 21*(9), 13–16. https://doi.org/10.1080/17415349.2006.11013505

Jones, C. T. (2021). "Wounds of regret": Critical reflections on competence, "professional intuition," and informed consent in research with intellectually disabled people. *Disability Studies Quarterly, 41*(2). https://doi.org/10.18061/dsq.v41i2.6869

Kuczewski, M. G. (2001). Disability: An agenda for bioethics. *American Journal of Bioethics, 1*(3), 36–44. https://doi.org/10.1162/152651601750418026

Kumar, R., & Lal, M. K. (2025). *Human growth and development*. Crown Publishing.

Lang, R., Kett, M., Groce, N., & Trani, J. F. (2011). Implementing the United Nations Convention on the rights of persons with disabilities: Principles, implications, practice and limitations. *Alter, 5*(3), 206–220. https://doi.org/10.1016/j.alter.2011.02.004

McCusker, P., Gillespie, L., Davidson, G., Vicary, S., & Stone, K. (2023). The United Nations convention on the rights of persons with disabilities and social work: Evidence for impact? *International Journal of Environmental Research and Public Health, 20*(20), 6927. https://doi.org/10.3390/ijerph20206927

McDaniel, K., & Friedman, M. (2021). Resisting the face of eugenics: Reclaiming portraiture through a disability gaze.

Mukherjee, D., Tarsney, P. S., & Kirschner, K. L. (2022). If not now, then when? Taking disability seriously in bioethics. *Hastings Center Report, 52*(3), 37–48. https://doi.org/10.1002/hast.1385

Olusanya, B. O., Kancherla, V., Shaheen, A., Ogbo, F. A., & Davis, A. C. (2022). Global and regional prevalence of disabilities among children and adolescents: Analysis of findings from global health databases. *Frontiers in Public Health, 10*, 977453. https://doi.org/10.3389/fpubh.2022.977453

Ovalle, A., Subramonian, A., Gautam, V., Gee, G., & Chang, K. W. (2023, August). Factoring the matrix of domination: A critical review and reimagination of intersectionality in AI fairness. In *Proceedings of the 2023 AAAI/ACM conference on AI, ethics, and society* (pp. 496–511). https://doi.org/10.48550/arXiv.2303.17555

Pezdek, K., & Rasiński, L. (2017). Between exclusion and emancipation: Foucault's ethics and disability. *Nursing Philosophy, 18*(2), e12131. https://doi.org/10.1111/nup.12131

Phillips, K. G., England, E., & Wishengrad, J. S. (2021). Disability-competence training influences health care providers' conceptualizations of disability: An evaluation study. *Disability and Health Journal, 14*(4), 101124. https://doi.org/10.1016/j.dhjo.2021.101124

Playford, R. C., Roberts, T., & Playford, E. D. (2015). Deontological and utilitarian ethics: A brief introduction in the context of disorders of consciousness. *Disability and Rehabilitation, 37*(21), 2006–2011. https://doi.org/10.3109/09638288.2014.989337

Rachmawati, M. R., Hasanbasri, M., & Hakimi, M. (2023). Virtue ethics among physicians who serve individuals with chronic spinal cord injury in Indonesia. *Asian Bioethics Review, 15*(3), 319–333. https://doi.org/10.1007/s41649-023-00245-6

Raposo, V. L. (2022). From public eugenics to private eugenics: What does the future hold? *JBRA Assisted Reproduction, 26*(4), 666–674. https://doi.org/10.5935/1518-0557.20220032

Remington, A. M. (2023). "Life Unworthy of Life" Aktion T4: The First Nazi Genocide. https://cupola.gettysburg.edu/student_scholarship/1063?utm_source=cupola.gettysburg.edu%2Fstudent_scholarship%2F1063&utm_medium=PDF&utm_campaign=PDFCoverPages

Retief, M., & Letšosa, R. (2018). Models of disability: A brief overview. *HTS Teologiese Studies/Theological Studies, 74*(1). https://doi.org/10.4102/hts.v74i1.4738

Roets, G., Remmery, M., Cautreels, D., Allemeersch, S., Benoot, T., & Roose, R. (2022). A critical exploration of institutional logics of de-institutionalisation in the field of disability policy and practice: Towards a socio-spatial professional orientation. *Social Work and Society, 20*(1).

Ruiz, F. J. (2025). Additive entanglement and intersectionality in UN human rights monitoring: Examining the inclusion of disability. *The International Journal of Human Rights, 29*(5), 842–863. https://doi.org/10.1080/13642987.2024.2430301

Saefudin, W., Putro, R. A., & Sriwiyanti, S. (2022). Restorative justice in child rape perpetrators: A case study on perpetrators with intellectual disability. *Jurnal Penelitian Hukum De Jure, 22*(1), 49–62. https://doi.org/10.30641/dejure.2022.V22.049-062

Savulescu, J. (2024). Two models of bioethics. *The American Journal of Bioethics, 24*(4), 37–38. https://doi.org/10.1080/15265161.2024.2312765

Serrato Calero, M. D. L. M., Delgado-Vázquez, Á. M., & Díaz Jiménez, R. M. (2021). Systematized review and meta-synthesis of the sterilization of women with disabilities in the field of social science: From macroeugenics to microeugenics. *Sexuality Research and Social Policy, 18*(3), 653–671. https://doi.org/10.1007/s13178-020-00488-0

Shaw, L. R., Chan, F., & McMahon, B. T. (2012). Intersectionality and disability harassment: The interactive effects of disability, race, age, and gender. *Rehabilitation Counseling Bulletin, 55*(2), 82–91. https://doi.org/10.1177/0034355211431167

Shikako, K., Lencucha, R., Hunt, M., Jodoin, S., Elsabbagh, M., Hudon, A., Cogburn, D., Chandra, A., Gignac-Eddy, A., Ananthamoorthy, N., & Martens, R. (2023). How did governments address the needs of people with disabilities during the COVID-19 pandemic? An analysis of 14 countries' policies based on the UN Convention on the Rights of persons with disabilities. *International Journal of Health Policy and Management, 12*, 7111. https://doi.org/10.34172/ijhpm.2023.7111

Sofokleous, R., & Stylianou, S. (2023). Effects of exposure to medical model and social model online constructions of disability on attitudes toward wheelchair users: Results from an online experiment. *Journal of Creative Communications, 18*(1), 61–78. https://doi.org/10.1177/09732586221136260

Steinert, C., Steinert, T., Flammer, E., & Jaeger, S. (2016). Impact of the UN convention on the rights of persons with disabilities (UN-CRPD) on mental health care research-a systematic review. *BMC Psychiatry, 16*(1), 166. https://doi.org/10.1186/s12888-016-0862-1

Thorneycroft, R. (2024). Screwing the social model of disability. *Scandinavian Journal of Disability Research, 26*(1). https://doi.org/10.16993/sjdr.1130

Tilley, E., Walmsley, J., Earle, S., & Atkinson, D. (2012). 'The silence is roaring': Sterilization, reproductive rights and women with intellectual disabilities. *Disability and Society, 27*(3), 413–426. https://doi.org/10.1080/09687599.2012.654991

Vanaken, G. J. (2022). Cripping vulnerability: A disability bioethics approach to the case of early autism interventions. *Tijdschrift voor genderstudies, 25*(1), 19–40. https://doi.org/10.5117/TVGN2022.1.002.VANA

Varkey, B. (2021). Principles of clinical ethics and their application to practice. *Medical Principles and Practice: International Journal of the Kuwait University, Health Science Centre, 30*(1), 17–28. https://doi.org/10.1159/000509119

Veerabathiran, R., & Thomas, S. M. (2025). Disability rights and policy in Asian and African continents. In *Disability across continents: Evolving policies and cultural shifts in Asia and Africa* (pp. 27–46). Springer Nature. https://doi.org/10.1007/978-981-96-6076-6_2

Veit, W., Anomaly, J., Agar, N., Singer, P., Fleischman, D. S., & Minerva, F. (2021). Can 'eugenics' be defended?. *Monash bioethics review, 39*(1), 60–67. https://doi.org/10.1007/s40592-021-00129-1

Watermeyer, B., & Swartz, L. (2023). Disability and the problem of lazy intersectionality. *Disability and Society, 38*(2), 362–366. https://doi.org/10.1080/09687599.2022.2130177

Weiss, M. (2024). *Apparitional representations: Disability history, reparative descriptions, and ethical failings in a special research collection.*

Wescott, H., MacLachlan, M., & Mannan, H. (2023). The use of a theory of change to promote knowledge management and global implementation of the United Nations Convention on the Rights of Persons with Disabilities (UNCRPD). *Disability inclusion and structural change: Understanding the relationship between stakeholders in the United Nations Partnership on the Rights of Persons with Disabilities (UNPRPD) programme*, 147. https://doi.org/10.47985/dcidj.397

WHO. (2022). https://www.who.int/publications/i/item/9789240063600

Wolbring, G. (2003). Disability rights approach toward bioethics? *Journal of Disability Policy Studies, 14*(3), 174–180. https://doi.org/10.1177/10442073030140030701

Yupanqui-Concha, A., Aranda-Farias, C., & Ferrer-Perez, V. A. (2021). Health practices of domination and exclusion: Views of activists, professionals and researchers on the situation of forced sterilization of women and girls with disabilities in Spain. *Saúde e Sociedade, 30*, e200107. https://doi.org/10.1590/S0104-12902021200107

Zaks, Z. (2024). Changing the medical model of disability to the normalization model of disability: Clarifying the past to create a new future direction. *Disability and Society, 39*(12), 3233–3260. https://doi.org/10.1080/09687599.2023.2255926

Chapter 2
Genetic Engineering, Screening, and Prenatal Ethics

Abstract The field of genetics has been growing with advancements in genomic technologies, such as CRISPR-based gene editing, gene therapy, and noninvasive prenatal testing (NIPT), which have revolutionized reproductive medicine, enabling early diagnosis and potential improvement of genetic disorders. Although these developments are helpful to humankind in diagnosing different diseases, they also cause profound ethical questions about autonomy, justice, and the social valuation in terms of disability. This chapter is about the intersection of genetic engineering, prenatal screening, and bioethics within the framework of disability studies. The chapter emphasizes the ethics of "curing" and "valuing" disability, the cultural assumption of genetic normalization, and the moral conflicts brought by disability-selective abortion through case studies involving genetic disorders. Additionally, focus on addressing counseling biases and the challenges of obtaining informed consent in the era of precision medicine, inclusion of disability rights, as well as equitable and thoughtful genetic practices, stabilize scientific progress with respect for multiculturalism, physical integrity, and reproductive choices.

Keywords Gene therapy · Informed consent · Noninvasive prenatal testing · Disability-selective abortion

2.1 Introduction

Biotechnology is swiftly going through an extraordinary transformation after the advent of genetic engineering. These innovations present significant societal, ethical, and legal challenges, despite their enormous potential to improve health and longevity. With its revolutionary solutions to complex biological problems, gene-editing technologies are transforming the fields of medicine and agriculture. It also offers fresh approaches to treating genetic problems, infections, and cancer (Wen et al., 2025).

Gene therapy, which comprises gene editing, gene substitution, and gene suppression, is used to cure diseases by altering genes. The therapeutics for genetic

A. Shibi Anilkumar, R. Veerabathiran, *Bioethics and Disability*, SpringerBriefs in Modern Perspectives on Disability Research,
https://doi.org/10.1007/978-981-95-8197-9_2

diseases have progressed remarkably in recent years due to gene therapy (Jiang et al., 2023). Better diagnostic precision, focused prevention, broader therapy possibilities, and the evolution of novel curatives have all been made possible by cutting-edge genome sequencing. In precision or individualized medicine, genetic information is used to ascertain disease susceptibility, prescribe therapy, and advise detection and prognosis. Investigation of the complex human genome, including genes that code for proteins and those that do not, is the focal point of genomic medicine, which has grown remarkably since the completion of the Human Genome Project (Anilkumar & Veerabathiran, 2025).

The expanding use of precision genotyping, which encompasses genetic testing and sequencing technologies, has enhanced prognostic results, aiding in the detection of the origin of genetic diseases for successful treatment, and has effectively reduced the rate of neonatal abnormalities. The prevalence of developmental disorders has decreased significantly as a result of advancements in medical research and technology (Hou & Zheng, 2024). Newer genomic techniques and technologies have expanded the scope and accelerated the development of novel diagnostic kits, vaccines, and treatment approaches, such as gene therapy and gene regulation (Pattan et al., 2021).

Genetics' emphasis shifted from its pediatric origins to its reproductive implications in the early 1970s. Families with higher-risk characteristics have been the primary focus of reproductive genetic testing. Numerous concerns are raised by the availability of presymptomatic testing, which involves examining individuals who are not yet exhibiting symptoms. The degree of social and economic assistance, as well as the degree of influence over life choices, may have a greater impact on the perceived quality of life for people with disabilities (PwD) than physical considerations (Kirschner et al., 2000).

The expanding field of precision medicine may strengthen and promote an individual's autonomy, choice, and successful management of their healthcare (De Paor & Blanck, 2016). Unconventional biotechnological discoveries and inadequate laws and regulations present the disability movement with complex moral, social, legal, and human rights issues. When misused, they hold the possibilities of bringing back historical eugenics; still, they also present chances to better the lives of PwD. Prospective threats can be brought down by careful national and international rules that strike a balance between knowledge gained from the past and long-term planning. When it comes to investigating or analyzing the moral and legal disputes that emerge from such biomedical progress, bioethics is indispensable (Conti, 2017).

Although gene editing technologies have extensive capabilities for ameliorating hereditary disorders, they also present significant ethical issues, particularly related to disability rights, as they aggravate ableism and societal stratification by promoting prejudice against PwD (Fatima et al., 2025). Modifications to procedures and rules ran the risk of violating women's right to reproductive autonomy (Johnston & Zacharias, 2017). This chapter explores the intersection of genetic engineering, screening, and prenatal ethics, aiming on how emerging genetic innovations challenge conventional notions of disability, humanity, and reproductive choice.

2.2 Ethical Perspectives on Genetic Conditions

2.2.1 *Genetic Screening and Diagnosis*

A large number of prenatal testing is designed for screening. These tests cover serum screening, carrier screening, and ultrasound. These tests aim to find women with pregnancies at high risk of congenital disabilities (Carlson & Vora, 2017). Newborn screening employs a range of modern analytical methods, including electrospray ionization tandem mass spectrometry (ESI-TMS) for concurrent detection of multiple metabolites, digital PCR, enzyme assays, high-performance liquid chromatography (HPLC), immunoassays, and next-generation sequencing (NGS) for specific identification of genetic mutations and inborn errors of metabolism (Remec et al., 2021). Figure 2.1 presents the methods used to identify genetic conditions.

For people with multiple congenital anomalies (MCA) and uncertain developmental delays (DD), genetic testing is part of the standard care. These conditions can be challenging to diagnose molecularly, as they are both medically and genetically distinct. Genetic testing is a standard practice for individuals with various

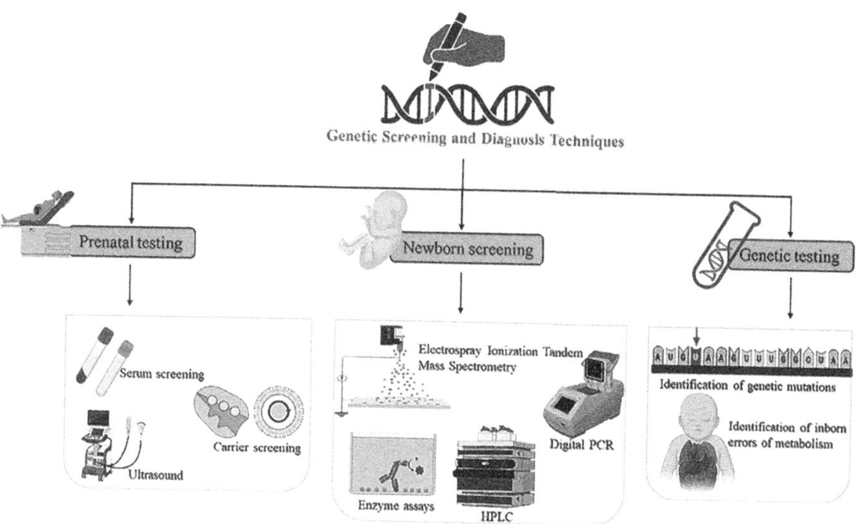

Fig. 2.1 Overview of genetic screening and diagnostic techniques. The figure illustrates three significant categories of genetic screening and diagnostic techniques: prenatal testing, newborn screening, and genetic testing. At the top, a stylized hand writing on a DNA helix symbolizes genetic analysis. Below this, three labelled sections contain visuals representing each testing type. The prenatal testing section includes icons of an ultrasound machine and blood sample tubes, as well as depictions of chorionic villus sampling and amniocentesis. The newborn screening section features laboratory equipment such as automated analyzers, mass spectrometers, and sample testing platforms. The genetic testing section features a DNA sequence, a marked mutation, and an illustration of an infant's internal organs, representing disease detection and molecular analysis

inherited abnormalities and unexplained developmental impairments (Li et al., 2021). Several experts have recognized that the application of genetic amenities directly needs a careful strategy based on ethics (Zhong et al., 2021).

2.2.1.1 Spinal Muscular Atrophy (SMA)

Degeneration of anterior horn cells in the spinal cord disrupts nerve signal communication between the brain and muscles, leading to spinal muscular atrophy (SMA), an autosomal recessive neuromuscular disorder. Muscle waste and weakening evolve as a result of this loss of transmission. Clinical appearance and severity of the illness vary (Boardman & Hale, 2018). A couple who have had a child with SMA has a 25% chance of having an affected child during each pregnancy, a 50% chance of having an asymptomatic carrier, and a 25% chance of having an unaffected child who is not a carrier (Prior et al., 2024).

Spinal muscular atrophy (SMA) exhibits a broad phenotypic spectrum classified into three types based on age of onset and motor milestones. In type II SMA, patients can sit but never walk, while those with type III achieve independent walking but experience varying levels of disability (Dunaway et al., 2012). Gradually, both type II and III patients may lose motor signs, with progressive deterioration in muscle strength and use (Feza Deymeer et al., 2008). On average, type III patients begin walking independently around 15 months but often lose this ability by 12 years; those requiring assisted walking typically lose it by age 7 (Russman et al., 1996).

Newborn screening (NBS) can identify spinal muscular atrophy (SMA), enabling prompt diagnosis and treatment before or shortly after symptoms begin. In one instance, a male baby with SMA was born at 32 weeks of gestation through spontaneous vaginal delivery after an early rupture of membranes and an otherwise uneventful pregnancy. Treatment was started before 40 weeks of age, as nusinersen has localized effects and has a decreased potential for systemic influence in premature newborns. The baby was given onasemnogene abeparvovec-xioi after receiving three loading doses of nusinersen at 36, 38, and 40 weeks of gestational age. No neurological abnormalities were found, and he tolerated nusinersen well (Nigro et al., 2023).

A novel gene therapy based on the onasemnogene abeparvovec (Zolgensma) was approved by the European Commission (EU) to treat patients with SMA 5q who have a bi-allelic mutation in the *SMN1* gene and a clinical diagnosis of SMA type 1, or who have a bi-allelic mutation in the *SMN1* gene and up to three copies of the *SMN2* gene. The recommended dosage applies to newborns and children with SMA up to a weight of 21 kg. Zolgensma is a once-in-a-lifetime gene therapy that replaces lost or malfunctioning *SMN1* gene function to address the disease's hereditary etiology (Ruta et al., 2024).

For the treatment of SMA, onasemnogene abeparvovec (OA) was approved by the FDA in May 2019 and the EMA in May 2020. Before approval, the Swiss Federal Social Insurance Office (FSIO) announced in July 2020 that newly diagnosed and treatment-naïve cases would be eligible for reimbursement of OA therapy under the Disability Insurance. Due to thrombotic microangiopathy and hepatic failure, adverse outcomes following OA treatment may be severe and, in rare instances, fatal (Stettner et al., 2023).

Another risdiplam functions as an *SMN2* splice modulator, stabilizing the ribonucleoprotein complex and competing with hnRNPG binding to promote exon 7 inclusion and the synthesis of full-length SMN proteins (Chaytow et al., 2021). Other treatments being researched focus on improving the survival and functionality of motor neurons and muscle through various methods, targeting pathways unrelated to the *SMN1* and *SMN2* genes. Combination therapy is being researched in light of the development and approval of several SMA disease-modifying treatments (Keinath et al., 2021).

In animal models, the combination of nusinersen with SMN-independent myostatin inhibitor therapy has demonstrated a favorable outcome that may be applicable to a different clinical strategy. New compounds are being developed as part of ongoing research to find more potential agents. All of the aforementioned therapies have the potential to be effective in treating SMA. However, the landscape of SMA is evolving rapidly due to new treatments that rely on a deeper understanding of SMA pathomechanisms (Chen, 2020).

Targeted transgene integration in both dividing and nondividing cells is made possible by the homology-independent targeted integration (HITI) approach, which is based on nonhomologous end joining (NHEJ). Homology-independent targeted integration (HITI) presents a promising strategy for the effective and long-lasting repair of SMA traits, as NHEJ is active throughout the cell cycle, including in neurons (Hatanaka et al., 2024). The cost burden for patients with SMA is estimated to be approximately 54 times higher than that of the healthy population (Paracha et al., 2022), and access to these therapies is limited in some countries (Armengol et al., 2024).

2.2.1.2 Down Syndrome

About 1 in 700 babies are born with Down syndrome, the most prevalent chromosomal condition (Chaiken et al., 2023). The most complicated and widespread genetic cause of intellectual disability is Down syndrome, the most common chromosomal condition caused by a third copy of full or part of chromosome 21 (HSA21) (Lorenzon et al., 2023). Among the most prevalent congenital conditions linked to intellectual disability is Down syndrome (DS). Significant speech and language deficits are a common feature of children with DS (Balambigai et al., 2023).

People with Down syndrome have a variety of physical health morbidities that differ among the population, and trisomy 21 is syndromic. A flat nasal bridge, upward-facing palpebral fissures, brachycephaly, low-set ears, and a single palmar crease are the characteristic physical characteristics of DS. Hirschsprung's disease, atlantoaxial joint instability, and atrioventricular septal anomalies are further related congenital abnormalities (Santoro et al., 2021).

Although a significant incidence is known to occur with increasing maternal age, the causes of this genetic change are mostly unknown (Huete-García & Otaola-Barranquero, 2021). For families with an affected child, the likelihood of a recurrence varies widely, ranging from 1% in most families to 100% in certain situations (Bull et al., 2022). Down syndrome (DS) is associated with an increased generation of mitochondrial ROS, which may contribute to immunosenescence through various distinct pathways. Adults with DS have aberrant CpG methylation in peripheral blood leukocytes and T lymphocytes that are specific to their genes (Illouz et al., 2021).

Chromosome 21 is duplicated, which causes Down syndrome (DS). By introducing the *XIST* gene, which is in charge of X-chromosome inactivation, into the extra chromosome 21 in induced pluripotent stem cells (iPSCs) from people with DS, a promising CRISPR-based method addresses this aberration. By efficiently silencing the additional chromosome, trisomic gene upregulation is reduced and gene expression returns to normal (Gondek et al., 2025).

In place of altering the gene sequence itself, a prominent method for modifying gene expression and activation may be to design and produce new epigenetic alterations that function as an on/off switch. Modern techniques such as chromosome--specific cDNA arrays have been utilized (Tafazoli et al., 2019). Gene therapy is employed to rectify or alter a mutation or damaged allele, such as the alteration of the survival motor neuron 2 (*SMN2*) gene transcript (Arabi et al., 2022). Biomarker--based therapeutic decision-making, as well as precautionary and combined interventions addressing comorbidities, may be utilized to inform next-generation treatments for Down syndrome (Lorenzon et al., 2023).

Prenatal diagnostic procedures were first developed in 1955. Numerous techniques for identifying chromosomal abnormalities have been developed since then. Because amniocentesis made it possible to diagnose Down syndrome (DS) accurately, healthcare systems quickly adopted it. Noninvasive prenatal testing (NIPT) methods have been developed more recently as safer substitutes for invasive procedures that do not carry the danger of fetal loss (Huete-García & Otaola--Barranquero, 2021).

Most nations now routinely offer prenatal screening for Down syndrome and other chromosomal disorders. Termination rates for prenatally diagnosed Down syndrome are typically high. A qualitative study of women's decisions to end a desired pregnancy due to Down syndrome revealed that the women were in a particularly ambiguous situation. Some women may face psychological repercussions in the short and long term, as well as persistent anxiety about whether the decision was right or wrong (Lou et al., 2021). Offering DS screening to encourage decisions that can result in selective abortion raises ethical concerns about fairness, social cohesion, and what it means to be healthy (Watson, 2021).

Supporting elderly persons with Down syndrome, specifically those experiencing dementia, has become more important due to a consequential shift in policy from institutionalization to community-based care. According to case studies, people with Down syndrome can still make noteworthy contributions to research even with their deteriorating abilities, emphasizing the requirement for inclusive models.

The significance of recognizing one's identity beyond a disability or dementia and the complication presented by unawareness about dementia were the two primary problems that surfaced. The reports urge decision-makers to enhance inclusion and communication in dementia care for individuals with disabilities (Watchman, 2016).

2.2.1.3 Deafness

Deafness is defined as a partial or total inability to hear. The four primary forms of deafness are conductive, sensorineural, mixed, and auditory neuropathy spectrum disorder. Conductive loss results from middle or outer ear blockage, sensorineural loss from inner ear or nerve damage, and mixed loss from both. The inner ear recognizes sound but is unable to transmit the information to the brain properly in auditory neuropathy. The range of hearing loss severity is profound (>91 dB) to slight (16–25 dB) (Kappaz, 2021).

A debilitating hearing impairment is typically defined as a persistent bilateral increase of the hearing threshold >40 decibels (dB) in the better ear (in those aged ≥15 years). In terms of frequency and impact, SNHI is the most common sensorineural impairment and a significant public health issue. About 70% of these inherited instances are caused by isolated (nonsyndromic) hearing impairment. Almost all of these types of deafness are monogenic. Severe to profound congenital deafness (or prelingual) primarily exhibits autosomal recessive inheritance (DFNB variants) (Petit et al., 2023).

Abnormalities in the inner ear or auditory nerve, such as harm to hair cells, auditory neurons, and supporting cells, result in sensorineural hearing loss (SNHL). Genetic material mutations are the leading cause of sensorineural hearing loss (Zhang et al., 2024). Genetic reasons account for almost half of the cases of congenital hearing loss. Due to these genetic variables, the illness may appear at birth or later in life, with varying degrees of severity (Hahn & Avraham, 2023).

The gene replacement method, frequently employed in inner ear gene therapy, replaces a faulty, mutant gene with a functional cDNA sequence that is delivered to targeted cells to restore normal gene function. It compensates for an inadequate single copy in dominant haploinsufficiency and provides a functional copy to two nonfunctional copies in recessive inheritance (Hahn & Avraham, 2023). In Beethoven (Bth) mice, a model of TMC1-related dominant hearing loss, gene suppression using RNA interference (RNAi) has been shown (Shibata et al., 2016). More recently, the highly selective CRISPR-Cas13 RNA editing technology has been used as a sophisticated RNA interference tool (Hahn & Avraham, 2023).

Cochlear implants and hearing aids are the main treatment options for inherited hearing loss. Now, there is more hope for treating congenital deafness due to gene therapies, such as gene replacement and gene editing, which repair or restore the expression of specific genes (Zhang et al., 2024). Adeno-associated virus (AAV)-mediated gene replacement therapy was used to partially restore hearing function in Vglut3-deficient mice in the first ear gene therapy trial, which began in 2012. Since then, gene therapy for AAV-mediated genetic deafness has advanced significantly, confirming its viability in several mouse models of deafness, including OtofΔ/Δ, Tmc1Y182C/Y182C, Tmc2Δ/Δ, StrcΔ/Δ, and Kcnq4W276S/+ (Akil et al., 2012).

The medical model views hearing loss as a physiological condition resulting from disorders of the auditory system. In clinical practice, inclusive approaches benefit both deaf children and healthcare providers by fostering positive environments. However, the deaf community continues to face significant health disparities due to exclusion from health surveillance, outreach programs, and public health messaging (Kappaz, 2021).

Deaf people are acknowledged as a language minority by science. Merging two conceptions of deafness either as a medical disability or as a cultural-linguistic identity, is a primary topic of debate in bioethical discussions. Although most of the participants admit deaf identities and sign languages as legitimate, they often give priority to the medical perspective, considering hearing loss as a physiological deficiency when deciding whether implantation is morally acceptable. Physical disability is perceived as more intrinsic than the social or linguistic aspects of deafness, replicating a blending of the medical and social conceptions of disability. The bioethical narratives about implants are shaped by the contradictions between viewing deafness as a physical impairment and as a cultural-linguistic identity (Kermit, 2009).

A case study involving carriers of the *Connexin 26* (*CX26*) gene, which is responsible for approximately half of all recessive hearing loss cases, used Preimplantation Genetic Diagnosis (PGD) to find genetic abnormalities in embryos produced through in vitro fertilization. Embryos without the defective gene were implanted or preserved for future use, while those carrying the inherited mutation were discarded. Bioethics rests on the belief in a shared "common morality" among all rational people. However, its interpretation varies: prescriptive common morality refers to moral principles that are normatively binding for everyone, while descriptive common morality reflects the moral practices commonly observed in society today (Lee et al., 2016).

2.3 The Ethics of "Curing" Versus "Valuing" Disability

To identify and treat disabilities like those above, genetic engineering and prenatal screening have become advanced. These methods frequently focus on characteristics that are classified as "abnormal" or "defective." The misuse of such genetic procedures could amount to a contemporary kind of eugenics, and viewing fetal impairments as making life "not worth living" carries the risk of applying the same judgment to those who already have similar disabilities (Conti, 2017).

Bioethical perspectives that strongly support the medical model are linked to "curing" or "eliminating" the harm caused by disabilities. A fundamental principle of bioethical research is the principle of beneficence (Conti, 2017). Additionally, the social model is predicated on several crucial distinctions between people with and without disabilities, the social and medical viewpoints, and impairment and disability (Shakespeare, 2006a).

The philosophy of cure, which is based on limited conceptions of what makes a meaningful or satisfactory life, is challenged by Critical Disability Studies (CDS). An assumed path from abnormality to normalcy that impacts medical discourse is reflected in the notion of a "curative imaginary." This viewpoint presents the elimination of disability as a universal and undeniable objective, suggesting that a therapy is always necessary, possible, and preferred. As a result, principles of normality and the perceived value of lives that differ from them become intertwined with beliefs about cure (Davis, 1995; Kafer, 2013).

The curative critical philosophical justifications are complicated and not always practical. It is frequently justified on three grounds: the desire to prolong life, the relief of suffering, and the advancement of equality of opportunity by lowering obstacles to involvement. There are, however, certain exceptions to each of these arguments, so treating a disability might not be the best or only approach to achieving these objectives (Stramondo, 2022).

Concepts such as disability pride and culture, once considered controversial, are now widely accepted (Gill, 2001). Many disabled individuals prioritize cultural and political goals over the pursuit of a cure. Studies show that individuals who believe health professionals largely determine their well-being—an attitude known as "health locus of control"—are more likely to accept the idea of a cure (Hahn & Belt, 2004).

The encouragement to view pride as the emotion to orient to and the "end" of the disability acceptance journey provides another affective expectation that is similar to developmentalist narratives in texts where disability pride is built on a rejection of overcoming narratives (Ingram & Jacobsen, 2025). According to the mere-difference perspective, disability is a variance in human experience rather than a fundamental weakness or disadvantage. It acknowledges that disabilities may result in the loss of particular commodities or abilities, frequently as a result of social treatment, but it also asserts that a disability opens doors to other special kinds of meaning and worth. Therefore, the value of disability is based on the idea that impairment is not just a lack but rather a different way of interacting with the environment (Barnes, 2014).

Rosemarie Garland-Thomson opposes the bioethical push toward genetic selection, contending that disability should be viewed as a valued human resource that should be preserved rather than just as a tragedy to be destroyed. She argues that the experience of disability creates narratives that unite and enhance the human community, and that interactions with nonnormative bodies enhance our understanding of humanity (Garland-Thomson, 2012).

According to Tom Shakespeare, even proponents of disability who disagree with the notion that impairments inevitably lower well-being are typically reluctant to lose any more of their own skills (Shakespeare, 2006b).

2.4 Disability Culture and the Ethics of Genetic Normalization

A shared language, a historical lineage that can be textually traced (through archives, memorials, and distinctive media/press publications), evidence of a cohesive social community, political solidarity, early acculturation within the family (and/or in segregated residential schools and clubs), genetic or generational ties, and pride and identity in segregation from others are all prerequisites for disabled people to claim a disability culture and, thus, a cultural identity (Peters, 2000).

A cultural perspective can also be used to comprehend disability, seeing it as a unique group identity apart from people without impairments. This viewpoint promotes a positive disability identity by highlighting pride in each person's distinct skills and qualities. Acknowledging disability as a cultural group opens doors for respect and inclusion in society, fostering the development of a strong and unified community (Hopson, 2019).

Disability culture fosters collective identity and consciousness that propel social and political movements by bringing people together through shared experiences, passions, and meanings. It promotes shared rituals, disability pride, and the growth of shared ideals and objectives. Disability culture, which has existed throughout history and spans all cultural boundaries, is distinct in that it can be involuntarily joined, frequently abruptly and unexpectedly (Hopson, 2019).

Identity and the body are often equated in ideas of normalcy. These notions of normalcy inform medical treatments that are performed on our bodies to make a person "healthy" and more "normal," such as straightening teeth and bones, treating illnesses, and addressing mental disorders. Many diseases, conditions, and behaviors are attributed to genetic information. Medical attempts to make people "healthy" and "normal" are motivated by ideas of normalcy, which connect identity to the body. Disability and difference are viewed as unwanted deviations that need to be rectified within this paradigm. This is reinforced by genetic knowledge, which promotes sameness as the foundation of equality and frames variance as a form of illness. In the past, these notions of normalcy allowed for moral judgment and societal control by portraying disability as a departure from a presumptive standard of physical competence (Taylor & Mykitiuk, 2001).

As genetic treatments progress, it will be necessary to define and regulate ethically acceptable techniques that take into account both explicit and implicit effects, such as how "enhancement" is evaluated in relation to normalcy norms. Since enhancement extends beyond restoring normal function to improving or surpassing typical human capacities, regulation necessitates careful evaluation of real-world impacts (Scully & Rehmann-Sutter, 2001).

The lack of disabled people's voices in medicine, incorrect presumptions about the quality of life for PwD, poor communication about genetic diagnoses, a narrow focus on medical complications, and a problematic attempt to eradicate not only genetic conditions but disabled people themselves have all been criticized by disability advocates in relation to various genetic technologies and the practice of genetic counseling. Workshops, advocacy groups, communication guidelines, and federal law have all been shaped by these criticisms (Houtz & Mueller, 2025).

Disability is frequently only explored in bioethics in relation to life-or-death choices, such as embryo selection, eligibility for medical help in dying, or triage amid resource shortages, while more comprehensive aspects of impaired lives are neglected. This emphasis on survival restricts ethical investigation and ignores the diversity and significance of impaired experiences. A more inclusive approach would look at how disability can influence new bioethical issues that go beyond accessibility or survival and question ableist notions of "quality of life." Furthermore, students with disabilities often complain about the lack of sufficient representation of disabilities in bioethics discussions, which usually rely on antiquated, ableist viewpoints. When combined with prejudice based on other identities, this lack of representation leads many people toward disability studies or away from university altogether (Dalrymple-Fraser, 2024; Waldschmidt, 2006).

2.5 Advances and Ethical Debates Surrounding Noninvasive Prenatal Testing (NIPT)

First discovered in 1988, noninvasive prenatal testing (NIPT) uses next-generation sequencing to analyze cell-free fetal DNA (cffDNA) from maternal serum. Because it provides an easy and low-risk way to identify prenatal abnormalities, governments and organizations worldwide are investing in its standardization and integration into healthcare systems. Health officials advise first-trimester screening to avoid difficulties since fetal aneuploidies are associated with congenital deformities (Priya, 2025).

Due to advancements in sequencing and bioinformatics, NIPT has expanded to screen for single-gene diseases, sex chromosomal aneuploidies (like Turner syndrome), and microdeletions (like DiGeorge syndrome). Uncertain conditions or fetal sex are examples of incidental findings that may emerge when testing grows more thorough. However, as genetic information increasingly influences decisions about maintaining a pregnancy, the greater breadth also raises ethical questions regarding selective abortion and the possibility of "designer babies" (Priya, 2025).

Noninvasive prenatal testing (NIPT) offers substantial benefits over conventional prenatal screening for Down syndrome, notably decreasing miscarriage risk by reducing the requirement for invasive tests. However, its ease and safety may make women feel pressured to accept testing, thereby confining their sense of choice. Women with a previous miscarriage often undergo increased anxiety during NIPT, affecting their decisions, perceptions of delay, and worries about future parenthood. Although decision-making autonomy is crucial, social and familial burdens are often tied to maternal age, which heavily influences choices about testing. Some women who gave birth post-NIPT stated feelings of insecurity about raising a child with a disability, indicating vague societal attitudes toward disability. Reinforcing welfare systems for PwD is thus important to maintain pregnant women's autonomy in NIPT decisions. These concerns extend beyond social impacts to incorporate personal, physical, and psychological dimensions (Yamamoto et al., 2022).

Ethical debates stress that the routinization of NIPT risks compromising informed and autonomous decision-making by standardizing testing and escalating social pressure on women. Researchers advise that repetitive use may lead to (a) unplanned choices made without any reflection, (b) restrained autonomy as testing becomes predictable, and (c) adverse effects on PwD through greater humiliation and reduced acceptance. From an ethical perspective, regularizing contests the principles of autonomy, informed consent, and respect for diversity. However, it is not always damaging if designed meticulously; routines can alternatively support informed, insightful decision-making rather than forced or automated testing (Rehmann-Sutter et al., 2023).

The situation of having too much information or the simplicity of the test, which restricts thorough pre-test counseling, can prevent independent decision-making in NIPT. To mitigate these concerns, a gradual disclosure model has been proposed that begins with general information and concludes with personalized discussions tailored to the woman's preferences. This approach not only benefits NIPT but also all prenatal testing situations by minimizing careless counseling, preventing information fatigue, and facilitating thoughtful, informed decisions (Schöne-Seifert & Junker, 2021).

High out-of-pocket expenses have led to concerns about unequal access to NIPT, especially for people from lower socioeconomic backgrounds who are affected by these costs the most. It is believed that equal access for all pregnant women would be the primary goal in the responsible use of NIPT, especially during public health campaigns. According to van der Meij et al. (2021), equitable access is considered a prerequisite for the moral and responsible use of NIPT in the context of a national prenatal screening program (van der Meij et al., 2021).

Nonetheless, the fact that several parents decide to abort their babies through genetic tests leads to worries that NIPT may influence society in such a way that it becomes more unacceptable to children with disabilities. Moreover, the discussion concerning the ethical usage of genetic technologies in reproductive choices is influenced by cultural variations (Jayashankar et al., 2023).

2.6 Ethical Implications of Disability-Selective Abortion

In the twentieth century, women's rights movements that advocated for the autonomy of women's reproductive rights and the "right to choose" mostly ignored disabled women. Disability often played a dual role as a reason for not granting abortion rights to disabled women and, at the same time, a reason for allowing abortions of fetuses with disabilities diagnosed (Jain & Sengupta, 2021). This situation mirrored a eugenic influence in state policies and legal discussions (Jain & Sengupta, 2021).

The discrepancies were confirmed again through the legal literature that was before the Medical Termination of Pregnancy (Amendment) Act, 2021, which allowed the disposal of pregnancies with fetal "abnormalities" and at the same time restricted the reproductive choices of others. The 2021 amendment has been received positively as a move toward ensuring dignity, autonomy, confidentiality, and justice for women who wish to undergo pregnancy abortion (Jain & Sengupta, 2021).

In the field of disability studies, disability-selected abortion (DSA) remains a topic of a very heated debate. Ontologically, DSA is an issue of significant concern, and it raises legal problems with cases of abortion where limitations on the gestation period do not apply to other pregnancy terminations (Ramaswamy, 2023). Even though the issues surrounding abortion remain hotly debated, discussions about abortion in the context of disabilities are not limited to the usual divide between "pro-choice" and "pro-life" positions. Disability can be considered as one of the causes for abortion without limiting the general accessibility of abortion. An expressivist objection would be sidestepped by a law that either prohibits all abortions or allows them for any reason up to birth, since disability would not be singled out as a unique justification (Herring & Robinson, 2024).

The interaction of political, social, and cultural elements about disability considerably influences the connection between selective abortion and prenatal diagnosis (PND). In such cases, women opting for abortion are often labeled "good mothers" since they think their decision is the safest for the fetus. From this perspective, the abortion is done for the child's good rather than as a hostile act toward the child. The ethical narrative presents the notion of the "appropriate aborter," where women sacrifice their emotional needs and personal convictions to save the fetus from a life considered to be full of hardships (Risøy & Sirnes, 2015).

The immunoregulatory skill of the expecting mother is crucial for a healthy pregnancy as the fetus carries the father's antigens (Varshini et al., 2022). Throughout pregnancy, routine check-ups enable doctors to identify potential cases that may require treatment and also help avoid over-treating low-risk cases, especially in women with established risk factors (Shibi Anilkumar et al., 2025).

2.7 Counseling Biases and the Principle of Informed Consent

Informed consent is a fundamental principle of medical ethics, which stipulates that individuals have the right to relevant, accurate, and unbiased information prior to receiving medical assistance, thereby empowering them to make informed treatment choices (Richardson & Nash, 2006). The concept of informed consent was first introduced by the Nuremberg Code, and its subsequent development was further outlined in the Declaration of Helsinki, the Belmont Report, and other organizational guidelines, including the European Society of Human Genetics regulations on direct-to-consumer testing. These principles have, however, been heavily utilized in all types of research involving human subjects outside of medical settings ever since (Budowle & Sajantila, 2023).

The foundation of informed consent comprises three essential elements: knowledge, comprehension, and free choice. Researchers are obliged to ensure that they present the consent materials clearly and concisely, covering the purposes, techniques, benefits, and risks of the study. Besides legal concepts such as due process and proportionality, ethical values like autonomy, justice, dignity, confidentiality, and solidarity, as well as democratic principles such as equality, transparency, and pluralism, are necessary for ensuring meaningful consent and indicate that this process should be constantly improved (Budowle & Sajantila, 2023).

The goal of informed consent is to get independent approval for acts that could otherwise infringe upon a person's rights. Legal and ethical consent for medical operations, such as genetic testing, must be obtained based on sufficient knowledge, competence, voluntariness, and comprehension. Clinical genetics has historically placed a strong emphasis on thorough, nondirective pre-test counseling, upholding the individual's right to know, or not know, genetic facts (Bunnik et al., 2013).

The person should be informed about the disorder, its risks and advantages, test limitations and reliability, the implications for family members, the likelihood of inheritance, and any accessible alternatives or support before giving their consent. However, for large-scale or genome-wide testing, when a lot of information can be overwhelming to comprehend, this meticulous procedure is often impractical. Parents or guardians act on behalf of minors because they are unable to give legal consent; however, their power is restricted if it violates the child's present or future rights, including the right to remain uninformed (Bunnik et al., 2013).

Nondirectiveness is when genetic counselors don't try to influence their clients' choices by giving them information and outlining potential options and results. This principle, founded on the idea that genetic testing decisions are personal, preserves client autonomy by ensuring that choices are made autonomously and in accordance with personal beliefs (National Society of Genetic Counselors, 2018; Chańska, 2022). A therapist's ability to truly connect with a patient may be hampered by biases that skew and anticipate the patient's emotions and worries (Yager et al., 2021).

The cultural and emotional factors of family expectations, religion, and social stigma exert considerable influence on genetic counseling and decision-making processes. For instance, family roles and cultural values might put communal peace above personal decision, while religious norms might give a moral basis that forms one's view of health and genetics, and stigma might cause nondisclosure or denial of genetic risks due to concerns about social and marriage prospects (Boardman et al., 2020; White, 2009). An inclusive counseling model that integrates unambiguous medical information and contributions from disability advocates and peer mentors fosters respect, empowerment, and trust by involving lived experiences and psychosocial support, thereby meeting the diverse needs of clients, which are not solely based on clinical facts (Pattison, 2005; Sakız et al., 2015).

2.8 Conclusion

Genetic manipulation and prenatal diagnosis are two significant developments in the modern biomedical field, one being a technological achievement and the other a moral challenge. Not only do these techniques promise to reduce pain and increase the number of choices in parenthood, but if they are applied without moral norms, there is a possibility of creating a society where people with certain genetic traits are favored over others, and the already existing unfairness is made worse. The chapter concludes that the responsible use of genomic technologies requires better counseling, transparent guidelines, and disability-inclusive ethics that prioritize dignity over perfection. A future development of biotechnology lies not in abolishing differences but in valuing them, ensuring that scientific innovation aligns with compassion, concern, human rights, and reverence for all humans.

References

Akil, O., Seal, R. P., Burke, K., Wang, C., Alemi, A., During, M., Edwards, R. H., & Lustig, L. R. (2012). Restoration of hearing in the VGLUT3 knockout mouse using virally mediated gene therapy. *Neuron, 75*(2), 283–293. https://doi.org/10.1016/j.neuron.2012.05.019

Anilkumar, A. S., & Veerabathiran, R. (2025). Genomics and congenital disabilities. In G. Bennett & E. Goodall (Eds.), *The Palgrave encyclopedia of disability*. Palgrave Macmillan. https://doi.org/10.1007/978-3-031-40858-8_510-1

Arabi, F., Mansouri, V., & Ahmadbeigi, N. (2022). Gene therapy clinical trials, where do we go? An overview. *Biomedicine and Pharmacotherapy, 153*, 113324. https://doi.org/10.1016/j.biopha.2022.113324

Armengol, V. D., Darras, B. T., Abulaban, A. A., Alshehri, A., Barisic, N., Ben-Omran, T., et al. (2024). Life-saving treatments for spinal muscular atrophy: Global access and availability. *Neurology: Clinical Practice, 14*(1), e200224. https://doi.org/10.1212/CPJ.0000000000200224

Balambigai, N., Chittathur, U. R., Kanagamuthu, P., & Ganesh, J. (2023). Investigating the use of gestures among children with down syndrome in India. *Journal of Modern Rehabilitation, 17*(4), 394–402. https://doi.org/10.18502/jmr.v17i4.13887

Barnes, E. (2014). Valuing disability, causing disability. *Ethics, 125*(1), 88–113. https://doi.org/10.1086/677021

Boardman, F. K., & Hale, R. (2018). How do genetically disabled adults view selective reproduction? Impairment, identity, and genetic screening. *Molecular Genetics and Genomic Medicine, 6*(6), 941–956. https://doi.org/10.1002/mgg3.463

Boardman, F. K., Clark, C., Jungkurth, E., & Young, P. J. (2020). Social and cultural influences on genetic screening programme acceptability: A mixed-methods study of the views of adults, carriers, and family members living with thalassemia in the UK. *Journal of Genetic Counseling, 29*(6), 1026–1040. https://doi.org/10.1002/jgc4.1231

Budowle, B., & Sajantila, A. (2023). Revisiting informed consent in forensic genomics in light of current technologies and the times. *International Journal of Legal Medicine, 137*(2), 551–565. https://doi.org/10.1007/s00414-023-02947-w

Bull, M. J., Trotter, T., Santoro, S. L., Christensen, C., Grout, R. W., & Council on Genetics. (2022). Health supervision for children and adolescents with Down syndrome. *Pediatrics, 149*(5), e2022057010. https://doi.org/10.1542/peds.2022-057010

Bunnik, E. M., de Jong, A., Nijsingh, N., & de Wert, G. M. (2013). The new genetics and informed consent: Differentiating choice to preserve autonomy. *Bioethics, 27*(6), 348–355. https://doi.org/10.1111/bioe.12030

Carlson, L. M., & Vora, N. L. (2017). Prenatal diagnosis: Screening and diagnostic tools. *Obstetrics and Gynecology Clinics of North America, 44*(2), 245–256. https://doi.org/10.1016/j.ogc.2017.02.004

Chaiken, S. R., Mandelbaum, A. D., Garg, B., Doshi, U., Packer, C. H., & Caughey, A. B. (2023). Association between rates of down syndrome diagnosis in states with vs without 20-week abortion bans from 2011 to 2018. *JAMA Network Open, 6*(3), e233684. https://doi.org/10.1001/jamanetworkopen.2023.3684

Chańska, W. (2022). The principle of nondirectiveness in genetic counseling. Different meanings and various postulates of normative nature. *Medicine, Health Care and Philosophy, 25*(3), 383–393. https://doi.org/10.1007/s11019-022-10085-0

Chaytow, H., Faller, K. M., Huang, Y. T., & Gillingwater, T. H. (2021). Spinal muscular atrophy: From approved therapies to future therapeutic targets for personalized medicine. *Cell Reports Medicine, 2*(7). https://doi.org/10.1016/j.xcrm.2021.100346

Chen, T. H. (2020). New and developing therapies in spinal muscular atrophy: From genotype to phenotype to treatment and where do we stand? *International Journal of Molecular Sciences, 21*, 9. https://doi.org/10.3390/ijms21093297

Conti, A. (2017). Drawing the line: Disability, genetic intervention and bioethics. *Laws, 6*(3), 9. https://doi.org/10.3390/laws6030009

Dalrymple-Fraser, C. (2024). Disabling bioethics futures. *Canadian Journal of Bioethics, 7*(1), 12–15. https://doi.org/10.7202/1110321ar

Davis, L. J. (1995). *Enforcing normalcy: Disability, deafness, and the body*. Verso.

De Paor, A., & Blanck, P. (2016). Precision medicine and advancing genetic technologies—Disability and human rights perspectives. *Laws, 5*(3), 36. https://doi.org/10.3390/laws5030036

Deymeer, F., Serdaroglu, P., Parman, Y., & Poda, M. (2008). Natural history of SMA IIIb: Muscle strength decreases in a predictable sequence and magnitude. *Neurology, 71*(9), 644–649. https://doi.org/10.1212/01.wnl.0000324623.89105.c4

Dunaway, S., Montes, J., Ryan, P. A., Montgomery, M., Sproule, D. M., & Vivo, D. C. D. (2012). Spinal muscular atrophy type III: Trying to understand subtle functional change over time—A case report. *Journal of Child Neurology, 27*(6), 779–785. https://doi.org/10.1177/0883073811425423

Fatima, A., Haider, A., & Batool, A. (2025). The human rights implications of scientific progress: A case study on gene editing and disability rights. *Journal of Engineering, Science and Technological Trends, 2*(1). https://doi.org/10.48112/jestt.v2i1.13

Garland-Thomson, R. (2012). The case for conserving disability. *Journal of Bioethical Inquiry, 9*(3), 339–355. https://doi.org/10.1007/s11673-012-9380-0

Gill, C. (2001). Divided understandings: The social experience of disability. In G. Albrecht, K. Seelman, & M. Bury (Eds.), *The handbook of disability studies* (pp. 351–372). Sage.

Gondek, K., Wojcieszek, M., Czarnota, M., Gacka, D., Fidelis, M., Żołnierek, A., et al. (2025). Gene therapy and down syndrome: Can the future of medicine bring a breakthrough? *Journal of Education, Health and Sport, 82*, 60525–60525. https://doi.org/10.12775/JEHS.2025.82.60525

Hahn, R., & Avraham, K. B. (2023). Gene therapy for inherited hearing loss: Updates and remaining challenges. *Audiology Research, 13*(6), 952–966. https://doi.org/10.3390/audiolres13060083

Hahn, H. D., & Belt, T. L. (2004). Disability identity and attitudes toward cure in a sample of disabled activists. *Journal of Health and Social Behavior, 45*(4), 453–464. https://doi.org/10.1177/002214650404500407

Hatanaka, F., Suzuki, K., Shojima, K., Yu, J., Takahashi, Y., Sakamoto, A., et al. (2024). Therapeutic strategy for spinal muscular atrophy by combining gene supplementation and genome editing. *Nature Communications, 15*(1), 6191. https://doi.org/10.1038/s41467-024-50095-5

Herring, J., & Robinson, H. (2024). A right to live without stigma? Examining negative stereotyping, negative messages, and Article 8 of the European Convention on Human Rights. *Legal Studies, 44*(4), 685–702. https://doi.org/10.1017/lst.2024.27

Hopson, J. (2019). Disability as culture. *Multicultural Education, 27*(1), 22–24.

Hou, K., & Zheng, X. (2024). A 10-year review on advancements in identifying and treating intellectual disability caused by genetic variations. *Genes, 15*(9), 1118. https://doi.org/10.3390/genes15091118

Houtz, C., & Mueller, R. (2025). Disability, genetic counseling, and medical education: From eugenics to anti-ableism. *Annual Review of Genomics and Human Genetics, 26*. https://doi.org/10.1146/annurev-genom-022024-010951

Huete-García, A., & Otaola-Barranquero, M. (2021). Demographic assessment of Down syndrome: A systematic review. *International Journal of Environmental Research and Public Health, 18*(1), 352. https://doi.org/10.3390/ijerph18010352

Illouz, T., Biragyn, A., Iulita, M. F., Flores-Aguilar, L., Dierssen, M., De Toma, I., et al. (2021). Immune dysregulation and the increased risk of complications and mortality following respiratory tract infections in adults with down syndrome. *Frontiers in Immunology, 12*, 621440. https://doi.org/10.3389/fimmu.2021.621440

Ingram, M., & Jacobsen, K. (2025). Both because of and in spite of: Towards the reclamation of queercrip joy. *Sexualities, 28*(3), 795–810. https://doi.org/10.1177/13634607241264319

Jain, D., & Sengupta, S. (2021). Reproductive rights and disability rights through an intersectional analysis. *Jindal Global Law Review, 12*(2), 337–357. https://doi.org/10.1007/s41020-021-00153-6

Jayashankar, S. S., Nasaruddin, M. L., Hassan, M. F., Dasrilsyah, R. A., Shafiee, M. N., Ismail, N. A. S., & Alias, E. (2023). Non-invasive prenatal testing (NIPT): Reliability, challenges, and future directions. *Diagnostics, 13*(15), 2570. https://doi.org/10.3390/diagnostics13152570

Jiang, L., Wang, D., He, Y., & Shu, Y. (2023). Advances in gene therapy hold promise for treating hereditary hearing loss. *Molecular Therapy, 31*(4), 934–950. https://doi.org/10.1016/j.ymthe.2023.02.001

Johnston, J., & Zacharias, R. L. (2017). The future of reproductive autonomy. *Hastings Center Report, 47*, S6–S11. https://doi.org/10.1002/hast.789

Kafer, A. (2013). *Feminist, queer, crip*. Indiana University Press.

Kappaz, S. (2021). *The significance of deaf culture: A bioethical analysis of pediatric cochlear implants*.

Keinath, M. C., Prior, D. E., & Prior, T. W. (2021). Spinal muscular atrophy: Mutations, testing, and clinical relevance. *The Application of Clinical Genetics*, 11–25. https://doi.org/10.2147/TACG.S239603.

Kermit, P. (2009). Deaf or deaf? Questioning alleged antinomies in the bioethical discourses on cochlear implantation and suggesting an alternative approach to d/Deafness. *Scandinavian Journal of Disability Research, 11*(2), 159–174. https://doi.org/10.1080/15017410902830744

Kirschner, K. L., Ormond, K. E., & Gill, C. J. (2000). The impact of genetic technologies on perceptions of disability. *Quality Management in Health Care, 8*(3), 19–26. https://doi.org/10.1097/00019514-200008030-00005

Lee, M., Chan, B., & Clark, P. A. (2016). Deafness and Prenatal Testing: A Case Study Analysis. *The Internet Journal of Family Practice, 14*(1). https://doi.org/10.5580/IJFP.39802

Li, C., Vandersluis, S., Holubowich, C., Ungar, W. J., Goh, E. S., Boycott, K. M., et al. (2021). Cost-effectiveness of genome-wide sequencing for unexplained developmental disabilities and multiple congenital anomalies. *Genetics in Medicine, 23*(3), 451–460. https://doi.org/10.1038/s41436-020-01012-w

Lorenzon, N., Musoles-Lleó, J., Turrisi, F., Gomis-González, M., De La Torre, R., & Dierssen, M. (2023). State-of-the-art therapy for Down syndrome. *Developmental Medicine and Child Neurology, 65*(7), 870–884. https://doi.org/10.1111/dmcn.15517

Lou, S., Hvidtjørn, D., Jørgensen, M. L., & Vogel, I. (2021). "I had to think: This is not a child." A qualitative exploration of how women/couples articulate their relation to the fetus/child following termination of a wanted pregnancy due to Down syndrome. *Sexual and Reproductive Healthcare, 28*, 100606. https://doi.org/10.1016/j.srhc.2021.100606

National Society of Genetic Counselors. (2018). National society of genetic counselors code of ethics. *Journal of Genetic Counseling, 27*(1), 6–8. https://doi.org/10.1007/s10897-017-0166-8

Nigro, E., Grunebaum, E., Kamath, B., Licht, C., Malcolmson, C., Jeewa, A., et al. (2023). Case report: A case of spinal muscular atrophy in a preterm infant: Risks and benefits of treatment. *Frontiers in Neurology, 14*, 1230889. https://doi.org/10.3389/fneur.2023.1230889

Paracha, N., Hudson, P., Mitchell, S., & Sutherland, C. S. (2022). Systematic literature review to assess the cost and resource use associated with spinal muscular atrophy management. *PharmacoEconomics, 40*(Suppl 1), 11–38. https://doi.org/10.1007/s40273-021-01105-7

Pattan, V., Kashyap, R., Bansal, V., Candula, N., Koritala, T., & Surani, S. (2021). Genomics in medicine: A new era in medicine. *World Journal of Methodology, 11*(5), 231–242. https://doi.org/10.5662/wjm.v11.i5.231

Pattison, S. (2005). Making a difference for young people with learning disabilities: A model for inclusive counselling practice. *Counselling and Psychotherapy Research, 5*(2), 120–130. https://doi.org/10.1080/17441690500258735

Peters, S. (2000). Is there a disability culture? A syncretisation of three possible world views. *Disability and Society, 15*(4), 583–601. https://doi.org/10.1080/09687590050058198

Petit, C., Bonnet, C., & Safieddine, S. (2023). Deafness: From genetic architecture to gene therapy. *Nature Reviews. Genetics, 24*, 665–686. https://doi.org/10.1038/s41576-023-00597-7

Prior, T. W., Leach, M. E., & Finanger, E. L. (2024). *Spinal muscular atrophy*. GeneReviews®[Internet].

Priya, P. D. (2025). Advancements in non-invasive prenatal testing: Implications for obstetric practice. *Scholar's Digest: Journal of Gynecology and Obstetrics, 1*(1), 1–16. https://scholarsdigest.org.in/index.php/sdjgo

Ramaswamy, K. (2023). *Legality of disability selective abortion: Discrimination and privacy laws under the ECHR*. https://doi.org/10.17863/CAM.111088.

Rehmann-Sutter, C., Timmermans, D. R., & Raz, A. (2023). Non-invasive prenatal testing (NIPT): Is routinization problematic? *BMC Medical Ethics, 24*(1), 87. https://doi.org/10.1186/s12910-023-00970-5

Remec, Z. I., Trebusak Podkrajsek, K., Repic Lampret, B., Kovac, J., Groselj, U., Tesovnik, T., et al. (2021). Next-generation sequencing in newborn screening: A review of current state. *Frontiers in Genetics, 12*, 662254. https://doi.org/10.3389/fgene.2021.662254

Richardson, C. T., & Nash, E. (2006). Misinformed consent: The medical accuracy of state-developed abortion counseling materials. *Guttmacher Policy Review, 9*(4), 6–12.

Risøy, S. M., & Sirnes, T. (2015). The decision: Relations to oneself, authority and vulnerability in the field of selective abortion. *BioSocieties, 10*(3), 317–340. https://doi.org/10.1057/biosoc.2014.39

Russman, B. S., Buncher, C. R., White, M., Samaha, F. J., & Iannaccone, S. T. (1996). Function changes in spinal muscular atrophy II and III. *Neurology, 47*(4), 973–976. https://doi.org/10.1212/WNL.47.4.973

Ruta, F., Ferrara, P., & Dal Mas, F. (2024). "No SMA can hold": Nursing care for children with spinal muscular atrophy. Descriptive analysis of two case studies. *Revista Científica de la Sociedad de Enfermería Neurológica (English ed.), 60*, 100144. https://doi.org/10.1016/j.sedeng.2024.100144

Sakız, H., Woods, C., Sart, H., Erşahin, Z., Aftab, R., Koç, N., & Sarıçam, H. (2015). The route to 'inclusive counselling': Counsellors' perceptions of disability inclusion in Turkey. *International Journal of Inclusive Education, 19*(3), 250–269. https://doi.org/10.1080/13603116.2014.929186

Santoro, J. D., Pagarkar, D., Chu, D. T., Rosso, M., Paulsen, K. C., Levitt, P., & Rafii, M. S. (2021). Neurologic complications of Down syndrome: A systematic review. *Journal of Neurology, 268*(12), 4495–4509. https://doi.org/10.1007/s00415-020-10179-w

Schöne-Seifert, B., & Junker, C. (2021). Making use of non-invasive prenatal testing (NIPT): Rethinking issues of routinization and pressure. *Journal of Perinatal Medicine, 49*(8), 959–964. https://doi.org/10.1515/jpm-2021-0236

Scully, J. L., & Rehmann-Sutter, C. (2001). When norms normalize: The case of genetic "Enhancement". *Human Gene Therapy, 12*(1), 87–95. https://doi.org/10.1089/104303401451004

Shakespeare, T. (2006a). The social model of disability. In *The disability studies reader* (pp. 16–24). Routledge.

Shakespeare, T. (2006b). *Disability rights and wrongs*. Routledge. https://doi.org/10.4324/9780203640098

Shibata, S. B., Ranum, P. T., Moteki, H., Pan, B., Goodwin, A. T., Goodman, S. S., et al. (2016). RNA interference prevents autosomal-dominant hearing loss. *The American Journal of Human Genetics, 98*(6), 1101–1113. https://doi.org/10.1016/j.ajhg.2016.03.028

Shibi Anilkumar, A., Mariam Thomas, S., & Veerabathiran, R. (2025). Autoimmune thyroid disease in pregnancy: Clinical implications for miscarriage risk. *Endocrine Research, 51*, 1–14. https://doi.org/10.1080/07435800.2025.2575170

Stettner, G. M., Hasselmann, O., Tscherter, A., Galiart, E., Jacquier, D., & Klein, A. (2023). Treatment of spinal muscular atrophy with Onasemnogene Abeparvovec in Switzerland: A prospective observational case series study. *BMC Neurology, 23*(1), 88. https://doi.org/10.1186/s12883-023-03133-6

Stramondo, J. A. (2022). A critique of the curative imperative. *Surgery, 171*(4), 1121–1122. https://doi.org/10.1016/j.surg.2021.10.008

Tafazoli, A., Behjati, F., Farhud, D. D., & Abbaszadegan, M. R. (2019). Combination of genetics and nanotechnology for down syndrome modification: A potential hypothesis and review of the literature. *Iranian Journal of Public Health, 48*(3), 371–378.

Taylor, K., & Mykitiuk, R. (2001). *Genetics, normalcy and disability*. ISUMA, Autumn.

van der Meij, K. R., Kooij, C., Bekker, M. N., Galjaard, R. J. H., Henneman, L., & Dutch NIPT Consortium. (2021). Non-invasive prenatal test uptake in socioeconomically disadvantaged neighborhoods. *Prenatal Diagnosis, 41*(11), 1395–1400. https://doi.org/10.1002/pd.6043

Varshini, G. S., Harshini, S., Siham, M. A., Tejaswini, G. K., Kumar, Y. S., Kulanthaivel, L., & Subbaraj, G. K. (2022). Investigation of FOXP3 (rs3761548) polymorphism with the risk of preeclampsia and recurrent spontaneous abortion: A systemic review and meta-analysis. *Asian Pacific Journal of Reproduction, 11*(3), 117–124. https://doi.org/10.4103/2305-0500.346089

Waldschmidt, A. (2006). Normalcy, bio-politics and disability: Some remarks on the German disability discourse. *Disability Studies Quarterly, 26*(2). https://doi.org/10.18061/dsq.v26i2.694

Watchman, K. (2016). Investigating the lived experience of people with down syndrome with dementia: Overcoming methodological and ethical challenges. *Journal of Policy and Practice in Intellectual Disabilities, 13*(2), 190–198. https://doi.org/10.1111/jppi.12167

Watson, L. (2021). Public health and ethics: The case of prenatal screening and Down's syndrome. *Journal of Public Health, 43*(3), e493–e494. https://doi.org/10.1093/pubmed/fdaa008

Wen, Z., Yang, D., Yang, Y., Hu, J., Parviainen, A., Chen, X., et al. (2025). The path to biotechnological singularity: Current breakthroughs and outlook. In *Biotechnology advances* (p. 108667). https://doi.org/10.1016/j.biotechadv.2025.108667

White, M. T. (2009). Making sense of genetic uncertainty: The role of religion and spirituality. *American Journal of Medical Genetics. Part C, Seminars in Medical Genetics, 151C*(1), 68–76. https://doi.org/10.1002/ajmg.c.30196

Yager, J., Kay, J., & Kelsay, K. (2021). Clinicians' cognitive and affective biases and the practice of psychotherapy. *American Journal of Psychotherapy, 74*(3), 119–126. https://doi.org/10.1176/appi.psychotherapy.20200025

Yamamoto, K., Chang, H., & Fukushima, A. (2022). Pregnant women's experiences of non--invasive prenatal testing (NIPT) in Japan: A qualitative study. *Journal of Genetic Counseling, 31*(2), 338–355. https://doi.org/10.1002/jgc4.1494

Zhang, L., Tan, F., Qi, J., Lu, Y., Wang, X., Yang, X., et al. (2024). AAV-mediated gene therapy for hereditary deafness: Progress and perspectives. *Advanced Science, 11*(47), 2402166. https://doi.org/10.1002/advs.202402166

Zhong, A., Darren, B., Loiseau, B., He, L. Q. B., Chang, T., Hill, J., & Dimaras, H. (2021). Ethical, social, and cultural issues related to clinical genetic testing and counseling in low-and middle--income countries: A systematic review. *Genetics in Medicine, 23*(12), 2270–2280. https://doi.org/10.1038/s41436-018-0090-9

Chapter 3
Disability Futures and Genetic Determinism and Posthumanism

Abstract This chapter examines how disability futures, genetic determinism, and posthumanism interconnect to reform present-day understandings of human disparity, embodiment, and enhancement. Improvements in genetics, gene therapy, and genome editing confront traditional philosophies of biological limits, while discussions over perfectionism and pluralism illuminate the ethical tensions between human excellence and the recognition of diversity. Disability advocacy reveals how experience, rights-based movements, and intersectional discrimination impact societal opinions of disability and flourishing. Posthumanist viewpoints further destabilize restrictions between the nonhuman and the human, the natural and the artificial, raising concerns about technological enhancement, inequality, and ableism. The chapter integrates philosophical, biomedical, and socio-political analyses to highlight how emerging technologies both transform and complicate disability futures.

Keywords Posthumanism · Genetic engineering · Enhancement · Transhumanism · Perfectionism

3.1 Introduction

Since genes were identified as the fundamental units of heredity, modifying the human genome has been a central goal of medicine. Although genes attract much attention, proteins perform most vital functions; genetic alterations can therefore disrupt protein activity and lead to disease (Al-Shammaa, 2024). Human evolution, shaped by complex patterns of variation, was once explained through multiregional models emphasizing geographic differences. However, morphological and genetic evidence since the late 1980s indicates that all humans are more closely related to one another than to any Late Quaternary hominin group (Mirazón Lahr, 2016).

Science has come a long way, and genetics not only helps to clarify but also to consider biological and medical phenomena. One of the major therapeutic strategies is gene therapy, which uses genetic engineering to treat diseases (Tamura & Toda,

A. Shibi Anilkumar, R. Veerabathiran, *Bioethics and Disability*, SpringerBriefs in Modern Perspectives on Disability Research,
https://doi.org/10.1007/978-981-95-8197-9_3

2020). Now, genome editing technologies have enabled the direct modification of genomic sequences in almost all eukaryotic cells, thereby allowing the creation of exact cellular and animal disease models and expanding applications from basic to biotech research (Li et al., 2020; Siham et al., 2023). Among recent discoveries, the greatest significant one has been CRISPR-Cas9, which has undeniably transformed our ability to edit genetic material with the highest precision by adding, removing, or altering mutations (Jinka et al., 2022; Youvan, 2024).

Together with these changes, pharmacogenomics (PGx), which studies the connections between genetic variation and drug response, has become a notable component of personalized medication and sustainable drug development by researching the role of genes in drug actions and reactions (Lakshmi et al., 2025). Concurrently, the combination of computational modeling and experimental validation has enabled a better study of medicinal target structure, function, and regulation (Das et al., 2022).

Genetics has had an effect not only on medical applications but also on the study of human behavior and social variances. Through twin and adoption studies, behavioral genetics has shown the extent to which specific traits, encompassing intellect, character, and psychopathology, are inherent, thereby opening an intense discussion over the ethical and social concerns of connecting genes with behavior (Harden, 2023). The growth of genetics has led to a confrontation with its outcomes across the spectrum of fields, from philosophy to sociology. Determinism in philosophy asserts that all events occur according to prescribed natural laws; genetic determinism extends this notion to the point that phenotypes are entirely the result of genotypes, with little or no influence from environmental or social factors (Harden, 2023; Hoefer, 2016).

The technology of genetic engineering captivates many people because it offers liberation from the limitations of human nature, the power to treat illnesses, eliminate congenital disorders, and enhance human characteristics. Among the supporters' dreams is a world of superior, comprehensive medical care, where machines could create people with higher IQs, greater muscularity, and more perfect body features, and at the same time society would not see any of these attributes as imperfections or disabilities. Transhumanists argue that if parents could choose their children's traits, few would willingly select blindness, paralysis, or other perceived limitations (Anchondo Pavón & Gallardo Macip, 2021).

Similarly, posthumanism suggests guiding human development through advanced technologies to create beings with few or no traits in common with modern humans. Through the modification or surpassing of human limitations, such novelties not only contest our notion of human life but also raise the question of whether society will be the ultimate winner or loser from this new enlightenment (Mendz & Cook, 2021).

In connection with this, posthuman disability studies is a new area that examines the relations between posthumanism and disability studies, analyzes the metamorphosing notions of the body, the redefinition of disability with respect to new technologies, and the changing understanding of humanity. These underlying conflicts between concepts play a critical role in the perception of disabled people (Romanska,

2024). Thus, this chapter investigates how Disability Futures, Genetic Determinism, and Posthumanism intersect to reshape our understanding of what it means to be human. It reflects on the ethical and philosophical consequences of altering or transcending human limitations through genetic and technological means, and how these ideas influence the value placed on disability, diversity, and human difference.

3.2 Philosophical Debate: Perfectionism Versus Pluralism

3.2.1 Conceptual Foundations of Perfectionism

There are at least two significant methods to understand what constitutes a decent human life. Such a life is interpreted in terms of well-being on the initial comprehension. A life that is as successful as possible for the individual leading it is the ideal life for a human being. According to the second interpretation, a person's good life is defined in terms of success or excellence. A great human life might be the finest in terms of well-being, but it doesn't have to be because such a life might include sacrificing one's own well-being for the benefit of other people or things. Therefore, the idea of a good human life is more expansive. Therefore, the concept of an exceptional human existence is more expansive than that of a life of great well-being. Furthermore, rather than using well-being, a generic description of perfectionism should use this broader concept (Wall, 2007).

The idea that what makes us human is the foundation for our goodness and that our ability to flourish as human beings depends on the quality of our most fundamental human traits is attributed to Aristotle. Hegel, Nietzsche, Aquinas, and Spinoza are all thought to have perspectives that could be interpreted as perfectionism. As a moral doctrine, perfectionism instructs people to defend and advance morally upright human existence (Bradford, 2017).

3.2.2 Dimensions and Models of Perfectionism

Perfectionism is a personality trait marked by an excessively high standard of performance and an unduly critical self-evaluation. From a psychopathological standpoint, Hollender was the first to define perfectionism as expecting oneself or others to perform at a level greater than what is necessary in a given circumstance (Hollender, 1965). The sources and objectives of perfectionistic impulses and behaviors vary across the three characteristics of perfectionism identified by Hewitt and Flett (Iranzo-Tatay et al., 2015).

The three-dimensional model of perfectionism, comprising other-oriented perfectionism (OOP; imposing high standards on others), self-oriented perfectionism (SOP; imposing high standards on oneself), and socially prescribed perfectionism (SPP;

perceiving that others expect perfection), is significant because it captures both intrapersonal (SOP) and interpersonal (OOP and SPP) aspects of perfectionism. This framework is extensively exercised as it presents a comprehensive method for understanding perfectionism, involving not only these core dimensions but also connected theories such as perfectionistic self-presentation and thought patterns (Hewitt et al., 2017).

The perfectionism studies have been guided by many conceptual models that have developed over time. Six aspects of perfectionism were identified by Frost et al. (1990): organization, parental expectations, parental criticism, personal standards, doubts about activities, and worry over mistakes. Hewitt and Flett (1991), on the other hand, defined perfectionism as self-oriented, socially dictated, and other-oriented. Lastly, Slaney et al. (2001) distinguished between order, disparity, and high standards (Burcaş & Creţu, 2021).

Since perfectionism has been connected to a variety of mental health conditions, it is regarded as one of the central psychological health disorders. A personality disposition known as perfectionism is characterized by the pursuit of excellence, extremely high performance standards, and excessively critical self-assessment. Both heredity and the environment influence perfectionism (Pourkhalili et al., 2022).

The genetic components of perfectionism have not been thoroughly studied. For instance, the Integrative Model of the Development of Perfectionism outlines three broad groups of elements and indicates multiple developmental routes for the emergence of this trait: child-specific elements (such as temperament and attachment style), external environmental elements (such as occupation, teachers, culture, and classmates), and parental elements (such as parenting methods and styles, parental personality, and objectives) (Oliveira et al., 2025).

According to the Social Expectations Model, parental expectations and criticism, along with contingent parental approval, can lead to the development of perfectionism. A link between parents and kids' perfectionism, as well as the impact of authoritarian parenting on kids' maladaptive perfectionism has been reported (Carmo et al., 2021).

3.2.3 *Perfectionism, Personality, and Disability*

Since personality traits influence attitudes toward disability, perfectionism may have a similar effect. Given its interpersonal nature, OOP, marked by critical, onerous expectations of others and limited empathy, may be most associated with negative attitudes toward people with disabilities (PwDs). Likewise, socially prescribed perfectionism (SPP) could contribute to such attitudes through broader difficulties in social interaction (Cox & Hill, 2018).

It is deemed "bad" news that must be shared with the parents if a newborn infant arrives who does not meet the standard of a "perfect" baby because of a physical difference or impairment. The dichotomous thinking about ability and disability is comparable to the binary thinking about perfection and imperfection. For instance, calling those without impairments "temporally-abled" suggests that a person loses all of their abilities once they have a condition. This disregards the fact that every human has a variety of skills and traits, some of which they may excel in. However, flaws and limitations will always coexist with traits perceived as close to perfection and skills (Martz, 2001).

People with disabilities (PwDs) have historically been marginalized by society because they are perceived as having a disability. The stigma and discrimination that PwDs face are a result of the idea that having a handicap signifies an innate state of being "less than others." Intersectional frameworks, which take a holistic view of personality, might help identify and comprehend strategies to reduce health disparities. People may be further excluded and shunned along other dimensions (such as socioeconomic class, race, or LGBTQIA+ status) if the numerous factors that contribute to the experience of disability are ignored. People who are disregarded as "deviant" yet are actually suffering from the effects of an ableist society that perpetuates injustice may be unfairly represented as a result of this absence (Brinkman et al., 2023).

3.2.4 Genetic Perfectionism and Enhancement Ethics

There are four categories of genetic engineering (Anderson, 1985). The first, somatic cell therapy, revises defective genes in somatic cells and involves only the treated individual, raising a few ethical issues. The second, germline therapy, amends reproductive cells, leaving changes to upcoming generations and prompting more ethical concern. The third type, enhancement, alters functional genes to progress traits such as intelligence or strength. Although restricted to the persons, it is extremely debated. The fourth, eugenic engineering, continues theoretical, encompassing the enhancement of gametes to produce genetically superior beings. With no clear boundaries between treatment and enhancement, the pursuit of perfection endangers, separating humanity into enhanced and unenhanced groups (Pugh, 2013).

Bioconservatives contradict the concept of perfection that foresees humans as masters of every facet of their lives through technology (Roduit et al., 2013). Sandel associates enhancement to a wish for controlling that oversees the gifted nature of human abilities and accomplishments. He advises that such enhancement efforts could direct to a domination of purposefulness over giftedness, control over respect, and manipulation over acceptance. In his view, the pursuit of enhancement technologies reflects human overestimation and presents a threat to critical virtues such as humbleness, solidarity, and responsibility, which are ultimate to living a good human life (Sandel, 2007).

Because of its connection with eugenics, the idea of enhancement conveys negative implications. Although the definition of eugenics has transformed over time, it is now meticulously linked to bias and the hierarchization of human life, as well as ableism, sexism, racism, and disability (Carapeto Raposo, 2022). Table 3.1 summarizes four major categories of genetic interventions along with the key ethical concerns associated with each approach.

Table 3.1 Categories of genetic engineering and their ethical implications

Category	Description	Ethical implications
Somatic cell therapy	Revises defective genes in somatic cells and involves only the treated individual	Raises few ethical issues
Germline therapy	Amends reproductive cells, leaving changes to upcoming generations	Prompts more ethical concern
Enhancement	Alters functional genes to progress traits such as intelligence or strength; restricted to the person	Extremely debated
Eugenic engineering	Theoretical; involves enhancement of gametes to produce genetically superior beings	Has no clear boundaries between treatment and enhancement; pursuit of perfection endangers society by separating humanity into enhanced and unenhanced groups

3.2.5 *Biomedical and Ethical Boundaries*

Biomedical enhancements, predominantly those involving genetic alteration, cause significant risks. Considering the level of well-being already achieved through past advancements, taking such risks is no longer justified. Thus, even if rejecting earlier forms of enhancement might once have been a mistake, it is now prudent to set clear limits (Van Niekerk, 2014).

Surrogacy has faced opposition due to the prevailing issue of social injustice, especially in the case of developing nations, where women, who are mostly illiterate and economically deprived, are likely to be misused and deprived of their rights to give proper consent. The criticisms also include the handling of women and children as commodities, the power over a surrogate's private life, the possibility of mental suffering, and the disputes with biological parents, besides the risk of a child being cast off if the baby is not up to the standards of "perfection" (Palazzani, 2020).

3.2.6 *Biological Normativity and Human Function*

The concept of biological normativity, which explains how biological processes could objectively malfunction without regard to human values, was proposed by Matthewson and Griffiths (2017) in response to criticisms of naturalism. Philosophical debates over health and illness frequently center on whether these concepts can be naturalized. The concepts of normality, abnormality, pathology, and physiology are essential for understanding biological function, even when human values are not involved and when illness is subjectively experienced as suffering or a barrier requiring medical aid (Veit, 2021).

3.2.7 *From Perfectionism to Biopluralism*

According to biopluralism, what unites all humans is that we each have unique bodies that influence, but do not define, how we relate to everyday reality. According to biopluralism, disability is a fundamental component of that pluralism, which does not necessitate that everyone consider themselves to be possibly impaired to support the assertion that disability is vital for democracy (McKinney, 2021).

Biostasis is the process of long-term human preservation with the expectation of future recovery, if possible. Two hypothetical modalities of biostasis can be distinguished: (a) provably reversible preservation and (b) preserving the body's informational features in a way that is not reversible with current technologies, with the hope that such technologies can be developed and implemented in the future. For humans, provably reversible preservation also referred to as suspended animation is not currently feasible and is unlikely to be achievable in the near future without extraordinarily rapid advances in preservation technology. The pursuit of biostasis raises some moral and social concerns that must be thoroughly investigated. Even though these social and ethical implications are necessary, these issues are intricate and complex, extending from resource allocation to feasible impacts on social institutions and individual freedom (McKenzie et al., 2024).

Progress optimists state that we are living more securely than past generations, while pessimists claim that life is getting inferior (Veenhoven, 2010). Enhancements aim to raise a desired capability within our species' normal range. Naturally, this definition raises important historical and ethical issues regarding embodiment, meaning, and purpose, such as the reasons for the desirability of specific traits or skills (Moeller & Andres Porras, 2025).

3.3 Disability Advocacy and Narratives of Flourishing

3.3.1 *Foundations of Advocacy and Social Justice*

Advocacy work is often contemplated as an essential part of nonprofit organizations' overall operations, as it enables them to engage with and epitomize their constituents. Examples of such initiatives include giving voice to miscellaneous viewpoints and demands, encouraging a more active civil society, reinforcing democracy and equality of opportunity, and fostering social and/or economic justice (Gelfgren et al., 2022). Minority groups have used advocacy as a social change strategy for decades to address power imbalances and demand equal rights (Flynn, 2013).

Speaking up for the rights of those who are being maltreated is the most fundamental definition of advocacy. Within the community and the legal system, advocacy can take several forms, such as representative or self-advocacy, systemic or individual advocacy, and informal or formal advocacy. The phrase "state-appointed" describes systems in which people with disabilities are assigned independent advocates to help them enforce their legal rights (such as access to healthcare, education, and social services) and ensure that their opinions are heard on issues that affect them. In addition to state-appointed sovereign advocates, it is crucial to recognize the role of numerous other community and voluntary forms of advocacy, such as peer, family, citizen, and self-advocacy, in guaranteeing the equal citizenship of PwDs (Flynn, 2013).

Among the guiding ideas that can be drawn from advocacy are the advancement of social justice and inclusiveness, the empowerment of PwDs, respect for autonomy, and the use of advocacy to promote systemic reform. All of these ideas, however, can be seen as supporting a central idea: the function of advocacy in guaranteeing good citizenship. People with disabilities (PwDs) have long desired equal citizenship (Flynn, 2010). Figure 3.1 illustrates how disability-related bioethics operates across three interconnected levels: healthcare practice, national governance, and global ethical frameworks, each contributing to the protection of rights, ethical standards, and equitable care for PwD.

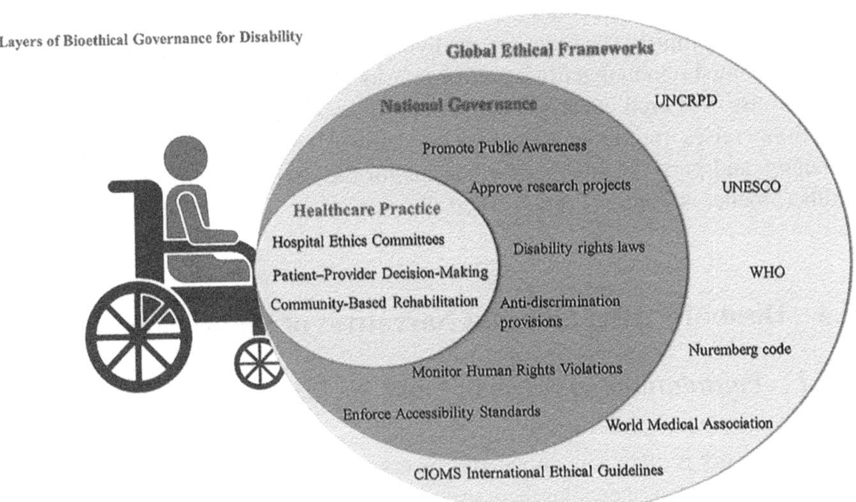

Fig. 3.1 Layers of bioethical governance for disability. The figure depicts three overlapping circles representing layers of bioethical governance for disability. The smallest yellow circle, labeled "Healthcare Practice," includes items such as hospital ethics committees, patient–provider decision-making, and community-based rehabilitation. This circle overlaps with a larger blue circle labeled "National Governance," which lists actions such as approving research projects, enacting disability rights laws, implementing anti-discrimination provisions, establishing accessibility standards, raising public awareness, and monitoring human rights violations. Both circles are encompassed by an even larger green circle titled "Global Ethical Frameworks," which includes institutions and guidelines such as the UNCRPD, UNESCO, WHO, the Nuremberg Code, the World Medical Association, and CIOMS international ethical guidelines. A red wheelchair-user icon appears on the left, symbolizing disability

3.3.2 Disability Movements, Power, and Representation

Disability studies scholars claim that although social exclusion and marginalization of people with psychiatric and nonpsychiatric disabilities manifest similarly, advocacy efforts, social movements, scholarship, and even healthcare benefits have addressed psychiatric and nonpsychiatric disabilities differently (Suarez-Balcazar et al., 2023). In addition to being a source of their dependence, welfare systems give disabled people one of the few chances for advocacy. In other words, while government support is essential for disability advocacy organizations, it may also diminish their influence. As a result, the disabled people's movement has made the connection between the government and organizations a top concern (Chang, 2017).

Disabled people have a better understanding of what it's like to have a disability than people without one. According to Brown (2016), activists with disabilities usually have a thorough awareness of the issues that affect them and individuals who share their interests; as individuals with impairments have a deeper understanding of what it's like to have a disability than those without. Individuals with disabilities who participate in activism or advocacy usually have a thorough understanding of the problems that affect them and others in similar circumstances. Policy discussions that disregard the perspectives of people with disabilities often result in laws that contain stigmatizing language, fail to address critical issues, or even limit services to those who need them (Hughes, 2016).

In the late 1960s, groups and conferences in Sweden and the US began to focus primarily on the experiences of people with learning disabilities, which is where self-advocacy initially emerged. Self-advocacy is one facet of the disability movement (Dowse, 2001). The autistic self-advocacy movement and some parent criticism have offered alternative viewpoints on how society should treat people with autism. According to others, this conflict is between those who are merely "mildly" disabled and those who are severely disabled (Ne'Eman & Bascom, 2020).

The goal of the disability rights movement has been to modify how the world is built so that everyone can engage in life's activities as fully and independently as possible. For many disabled individuals, the struggle for social justice is more than just a political issue; it's a lived philosophy that represents years of hardship and personal development. As a result of this development, struggle, and learning, it becomes evident that disabled people have much more in common than just physical differences; they are oppressed. In pursuit of justice and equal rights, the Disability Rights Movement is pushing society to amend how it treats disabled individuals (Loewen & Pollard, 2010).

3.3.3 International Frameworks and Rights-Based Advocacy

The UN General Assembly enacted the Convention on the Rights of PwDs in December 2006, and as of 2017, 160 governments had ratified it. It asserted that PwDs should be viewed as "subjects" with rights who may exercise their rights, make life decisions based on their free and informed consent, and actively engage in society rather than as "objects" of charity, medical care, and social protection (United Nations, 2017).

Although nations across the globe support the United Nations Convention on the Rights of PwDs (2006), when faced with conflicting economic goals, their policies and practices too often severely harm children with disabilities and their families. Researchers have established theoretical frameworks to assist policy creation and the evaluation of activities targeted at improving the results of PwDs. A helpful framework for assessing an individual's quality of life in a number of areas is the Quality of Life (QoL) concept. Notably, it has been demonstrated that the concept of QoL is positively related to the UN Convention's Articles. Disability is typically a racialized, gendered, sexed, ethnicized, and class-based experience that operates within a framework of complex and multidimensional patterns of identities and inequities (Sauer & Lalvani, 2017).

3.3.4 Narratives, Experience, and Flourishing

Flourishing extends beyond physical wellness, encompassing psychological, social, cultural, moral, existential, and spiritual dimensions. It involves confronting hardship and limitation, not just seeking pleasure. As a phenomenological concept, flourishing focuses on the lived experience of how individuals interpret and construct meaning within social and cultural contexts. In rehabilitation, this approach emphasizes understanding what "functioning well," "impairment," or "success" mean for each person, as shaped by their unique moral, relational, and life-world perspectives (Haque et al., 2020).

One method of bringing the past into the present and allowing for fresh perspectives on what was experienced is through narrative. The idea of Narrative-Based Medicine (NBM) arose in the 1990s in response to the changes in medical practice brought about by Evidence-Based Medicine. Narrative-Based Medicine (NBM) addresses the significance of the patient's experience of becoming ill and the clinical encounter for the establishment of appropriate healthcare, based on an interpretive paradigm (Vieira & Favoreto, 2016).

3.3.5 *Advocacy, Development, and Global Sustainability*

Environmental justice and all marginalized groups, especially PwDs, must be taken into account when transforming our societies toward sustainability. Multifaceted obstacles to inclusion and involvement in decision-making are frequently historical and lead to poverty and lower levels of information and resources, as well as decreased access to education, health care, and jobs. It may be possible to transcend the existing unequal power dynamics in sustainable decision-making by taking intersectionality into account when making decisions for impaired populations (Kosanic et al., 2022).

Disabled communities should be actively involved in international sustainability research initiatives like Future Earth and science-policy interaction platforms that promote sustainability, like IPBES and IPCC. We cannot create inclusive risk management and adaptation plans, inclusive conservation policies, or sustainable, just futures unless we include and involve disabled communities as pertinent stakeholders (Kosanic et al., 2022).

Another developmental concern is disability. Poverty and disability have a reciprocal relationship, necessitating contextual knowledge and assistance. People with disabilities (PwDs) face social marginalization and disempowerment, not only a lack of material resources, according to Amartya Sen's capabilities approach, which goes beyond the concept of usefulness (such as gross domestic product) to highlight human rights and "development as freedom"(Sen, 2000).

The organizing system of the UN Convention argues that PwDs have equal rights to social protection, and that the state should provide safety nets to help them live empowered lives (Guide, 2014). Disability is referenced in many Sustainable Development Goals, specifically those related to inequality, education, data collection and monitoring, accessibility of human settlements, and growth and employment. Many international organizations patronage community-based rehabilitation (Samuel & Jacob, 2018).

3.4 Posthumanism

As Iovino notes, the foundations of modern posthumanism were laid in the late twentieth century, with Haraway's "A Cyborg Manifesto" (1985) as a defining early contribution (Iovino, 2016). Haraway identifies three major boundary breakdowns: (1) the human–animal divide, breached as animals demonstrably share language, tool use, social behavior, and mental events; (2) the organism–machine divide, as late-twentieth-century technologies blur distinctions between self-developing/externally designed systems, and natural/artificial, mind/body; and (3) the physical–nonphysical divide, since for "cyborgs," thought, information, and consciousness exist on the same ontological continuum as matter (Hicks, 2022).

Posthumanism is an "umbrella" term that refers to a range of methods and ideologies that share a rejection of humanism rather than a single, cohesive notion. The various dichotomies that surround Western conceptualizations, such as mind/body, human/nonhuman, subject/object, or nature/culture, have been challenged by posthumanist methods. It has been said that these divisions enable the creation of political tactics with deadly repercussions for people on the "erroneous" side of the line (De Liaño & Fernández-Götz, 2021).

3.4.1 Posthumanism Versus Transhumanism

Transhumanism and posthumanism are two distinct schools of thought that are fundamentally at odds with one another. Transhumanism concerns man, his inherent limits, and possible ways to overcome them. Posthumanism, in turn, widens the very realm of agency by incorporating nonhuman objects and nonhuman beings, and by challenging the binary obstructions of culture–nature, humanism–antihumanism, or human–nonhuman. The repudiation of biological and ethical humanity is the ideological basis of posthumanism. (Merzlyakov, 2022).

3.4.2 Technology, Nonhuman Agency, and Anthropocentrism

Modern posthumanist conceptions of technology critically rely on nonhuman elements, such as computers, actively contributing to their own development rather than serving as passive platforms for human advancement. Recognizing assertive or sovereignty qualities in nonhumans, including robots, has turned out to be essential to posthumanists' disquisition of the binary categories that underlie post-Enlightenment thought, such as human/nonhuman and animate/inanimate, anthropocentrism and humanism (Cadman et al., 2025).

3.4.3 Posthumanism's Critique of Enlightenment Humanism

The belief that humans occupy a privileged position characterized by rationality, autonomy, and individualism is contested by posthumanism, which emerged from critiques of Enlightenment humanism. By acknowledging the agency of nonhuman creatures like animals, robots, ecosystems, and artificial intelligence, it modernizes the human. It contends that distinctions between human and nonhuman, natural and artificial, and biological and technical are becoming unclear in a time of biotechnology and digital progress. Posthumanism challenges long-standing anthropocentric presumptions in philosophy, science, and ethics and behests a reconsideration of what it means to be human in the face of technological advancement, globalization, and planetary catastrophes (Ferrando, 2019).

3.4.4 Diversity Within Posthuman Theory

There are numerous branches and disagreements among various methods within posthuman theory, which is not a single concept. Studying posthuman disability could help us "think again about ourselves, our relations, and our politics (Saur & Sidorkin, 2018).

3.4.5 Posthumanism and the Critique of the "Ideal Human"

Braidotti (2013) argues that the Western ideal of "the human" emerged as a healthy, rational, White male, an allegedly universal model of bodily perfection. Posthumanism challenges this closed, transcendental notion, highlighting how people with impairments have been cast as "abnormal," "not fully human," or the "Other" (Braidotti, 2013). The category of "disabled" lies on the myth that traits like ability, gender, or normality are fixed at birth, when these boundaries are socially constructed and exclusionary. As an umbrella term, *the posthuman* includes genetically modified persons, organic artificial hybrids, chimeras, and cyborgs not a fixed identity but an expanded, plural understanding of the human. By decentering the humanistic ideal of superiority over other beings, posthumanism offers disability studies a valuable framework for rethinking what counts as human and for embracing diversity and plurality (Gibson, 2015).

3.4.6 Posthuman Disability Studies and Technological Enhancement

Posthuman disability studies focusses on the interdisciplinary intersections of posthumanism and disability studies, including the evolving concept of body and bodily identity, the evolving definition of disability in the context of emerging technologies, and changes in human perceptions. Digital and artificial intelligence technologies' prosthetic features, along with posthuman biomimicry, assume that the human body is somehow disabled: a body that is not connected to any machine is an inferior body (Romanska, 2024).

Reconstructive surgery, nootropics, prescription stimulants, and other technological innovations that were initially developed to "fix" the disabled body have been adopted by nondisabled populations as ways to improve the abilities of the nondisabled body (plastic surgery, performance-enhancing use of prescription drugs, for example). For example, exoskeleton suits, which were originally designed for PwDs, are currently being studied to "supercharge human workers." By utilizing AI tools to improve both the AI's and their own capabilities, the disabled are building and exploring a symbiotic relationship between AI and humans that could impact how we engage with and govern AI technology in the future (Romanska, 2024).

3.4.7 *Enhancement Ethics and Social Inequality*

Human rights and equality in a future society are challenged by the potential and justification of genetic modifications (as well as other forms of enhancement, such as implants, prostheses, and transgenesis) that increase human capacities. Beyond that, steps must be taken to avoid inequity between "normal" (just human) and "augmented" (posthuman) humans in posthuman society. Since most situations fall within the posthumanist range, improvements could lead to significant social inequality. Technology hasn't developed to the point where posthuman individuals with intelligence or skills much above the typical human range could exist (Triviño, 2015).

3.5 Conclusion

Redefining humanity is a complex and multidimensional process that goes along with culture, ethics, and politics and is connected to concepts such as disability futures, genetic determinism, and posthumanism. Although genetic enhancement and technology offer human beings the power to interfere and improve to a step never seen before, they also create a major threat of creating new systems of discrimination, perfectionism, and inequality. The activists for people with disabilities, along with the philosophers who claim human flourishing, remind us that human value cannot be reduced to biological traits or technological capability, but is determined by experience, relations, and the various ways of being. Posthumanism is a crucial theory that can be used to question the hierarchical binaries of normal/abnormal and enabled/disabled. Yet, it still raises the necessity for careful ethical considerations, particularly in the areas of enhancement and social justice. As societies are being converted into technology-mediated ones, the necessity of pluralism, autonomy, rights, and equitable participation as the core priorities of the future is obvious. Thus, to achieve real progress, it is not necessary to strive for perfection but rather to develop a future that is inclusive and honors human diversity in all its aspects.

References

Al-Shammaa, N. F. K. (2024). Advancing medicine: The potential of gene therapy to treat genetic disorders. *Genetics and Molecular Research, 23*(4), 1–8. https://doi.org/10.4238/gmr2378

Anchondo Pavón, S., & Gallardo Macip, C. (2021). The perfection of vulnerable lives. Genetic modification and disability. *Medicina y ética, 32*(2), 483–552. https://doi.org/10.36105/mye.2021v32n2.04

Anderson, W. F. (1985). Human gene therapy: Scientific and ethical considerations. *The Journal of Medicine and Philosophy, 10*(3), 275–292. https://doi.org/10.1093/jmp/10.3.275

Bradford, G. (2017). Problems for perfectionism. *Utilitas, 29*(3), 344–364. https://doi.org/10.1017/S0953820816000418

Braidotti, R. (2013). *The posthuman* (p. 192). Polity Press.

Brinkman, A. H., Rea-Sandin, G., Lund, E. M., Fitzpatrick, O. M., Gusman, M. S., Boness, C. L., & Scholars for Elevating Equity and Diversity (SEED). (2023). Shifting the discourse on disability: Moving to an inclusive, intersectional focus. *The American Journal of Orthopsychiatry, 93*(1), 50–62. https://doi.org/10.1037/ort0000653

Brown, E. (2016). Disability awareness: The fight for accessibility. *Nature, 532*(7597), 137–139. https://doi.org/10.1038/nj7597-137a

Burcaş, S., & Creţu, R. Z. (2021). Perfectionism and neuroticism: Evidence for a common genetic and environmental etiology. *Journal of Personality, 89*(4), 819–830. https://doi.org/10.1111/jopy.12617

Cadman, S., Tanner, C., & Pang, P. C. I. (2025). Humanism strikes back? A posthumanist reckoning with 'self-development' and generative AI. *AI & Society*. https://doi.org/10.1007/s00146-025-02339-1

Carapeto Raposo, V. L. (2022). The better I can be: In defence of human enhancement for a new genetic equality. *Canadian Journal of Bioethics, 5*(2), 189–202. https://doi.org/10.7202/1089801ar

Carmo, C., Oliveira, D., Brás, M., & Faísca, L. (2021). The influence of parental perfectionism and parenting styles on child perfectionism. *Children, 8*(9), 777. https://doi.org/10.3390/children8090777

Chang, H. H. (2017). From advocacy to service provision: State transformation and the disability rights movement in Taiwan. *Disability & Society, 32*(3), 308–322. https://doi.org/10.1080/09687599.2017.1294050

Cox, N. C., & Hill, A. P. (2018). Trait perfectionism and attitudes towards people with disabilities. *Personality and Individual Differences, 122*, 184–189. https://doi.org/10.1016/j.paid.2017.10.028

Das, D., Bihari Jena, A., Banerjee, A., Kumar Radhakrishnan, A., Duttaroy, A. K., & Pathak, S. (2022). Can plant-derived anti-HIV compounds be used in COVID-19 cases? *Medical Hypotheses, 166*, 110926. https://doi.org/10.1016/j.mehy.2022.110926

De Liaño, G. D., & Fernández-Götz, M. (2021). Posthumanism, new humanism and beyond. *Cambridge Archaeological Journal, 31*(3), 543–549. https://doi.org/10.1017/S095977432100024X

Dowse, L. (2001). Contesting practices, challenging codes: Self advocacy, disability politics and the social model. *Disability & Society, 16*(1), 123–141. https://doi.org/10.1080/713662036

Ferrando, F. (2019). *Philosophical Posthumanism*. https://www.torrossa.com/it/resources/an/5210722

Flynn, E. (2010). A socio-legal analysis of advocacy for people with disabilities—Competing concepts of 'best interests' and empowerment in legislation and policy on statutory advocacy services. *Journal of Social Welfare & Family Law, 32*(1), 23–36. https://doi.org/10.1080/09649061003675840

Flynn, E. (2013). Making human rights meaningful for people with disabilities: Advocacy, access to justice and equality before the law. *The International Journal of Human Rights, 17*(4), 491–510. https://doi.org/10.1080/13642987.2013.782858

Frost, R. O., Marten, P., Lahart, C., & Rosenblate, R. (1990). The dimensions of perfectionism. *Cognitive Therapy and Research, 14*(5), 449–468. https://doi.org/10.1007/BF01172967

Gelfgren, S., Ineland, J., & Cocq, C. (2022). Social media and disability advocacy organizations: Caught between hopes and realities. *Disability & Society, 37*(7), 1085–1106. https://doi.org/10.1080/09687599.2020.1867069

Gibson, H. (2015). Exploring contemporary anthropological theory: Can we use posthumanism to reconceptualise the disabled body? *Sites: A Journal of Social Anthropology & Cultural Studies, 12*(2). https://doi.org/10.11157/sites-vol12iss2id297

Guide, T. (2014). *The convention on the rights of persons with disabilities*. United Nations of Human Rights.

Harden, K. P. (2023). Genetic determinism, essentialism and reductionism: Semantic clarity for contested science. *Nature Reviews Genetics, 24*(3), 197–204. https://doi.org/10.1038/s41576-022-00537-x

Haque, O. S., Lenfest, Y., & Peteet, J. R. (2020). From disability to human flourishing: How fourth wave psychotherapies can help to reimagine rehabilitation and medicine as a whole. *Disability and Rehabilitation, 42*(11), 1511–1517. https://doi.org/10.1080/09638288.2019.1602674

Hewitt, P. L., & Flett, G. L. (1991). Perfectionism in the self and social contexts: Conceptualization, assessment, and association with psychopathology. *Journal of personality and social psychology, 60*(3), 456–470. https://doi.org/10.1037//0022-3514.60.3.456

Hewitt, P. L., Flett, G. L., & Mikail, S. F. (2017). *Perfectionism: A relational approach to conceptualization, assessment, and treatment*. Guilford Press.

Hicks, A. J. (2022). Posthumanism. In *The encyclopedia of contemporary American fiction 1980–2020* (Vol. 2, pp. 1–10). https://doi.org/10.1002/9781119431732.ecaf0258

Hoefer, C. (2016). Causal determinism. In E. N. Zalta (Ed.), *The Stanford encyclopedia of philosophy (Spring 2016 edition)*. https://plato.stanford.edu/archives/spr2016/entries/determinism-causal

Hollender, M. H. (1965). Perfectionism. *Comprehensive Psychiatry, 6*(2), 94–103. https://doi.org/10.1016/S0010-440X(65)80016-5

Hughes, J. M. (2016). Increasing neurodiversity in disability and social justice advocacy groups. *Autistic Self Advocacy Network*, 1–21.

Iovino, S. (2016). Posthumanism in literature and ecocriticism. *Relations: Beyond Anthropocentrism, 4*(1), 11–22.

Iranzo-Tatay, C., Gimeno-Clemente, N., Barberá-Fons, M., Rodriguez-Campayo, M. Á., Rojo-Bofill, L., Livianos-Aldana, L., et al. (2015). Genetic and environmental contributions to perfectionism and its common factors. *Psychiatry Research, 230*(3), 932–939. https://doi.org/10.1016/j.psychres.2015.11.020

Jinka, C., Sainath, C., Babu, S., Chennupati, A. C., Muppidi, L. P., Krishnan, M., Sekar, G., Chinnaiyan, M., & Andugula, S. K. (2022). CRISPR-Cas9 gene editing and human diseases. *Bioinformation, 18*(11), 1081–1086. https://doi.org/10.6026/973206300181081

Kosanic, A., Petzold, J., Martín-López, B., & Razanajatovo, M. (2022). An inclusive future: Disabled populations in the context of climate and environmental change. *Current Opinion in Environmental Sustainability, 55*, 101159. https://doi.org/10.1016/j.cosust.2022.101159

Lakshmi, K., Nagarajan, B., Dabburu, K., Roy, C., Shanthi, B., Das, G., et al. (2025). Pharmacogenomics for sustainable drug development: A narrative review of precision medicine, green chemistry, and multi-omics innovation. *Journal of Applied Pharmaceutical Science*. https://doi.org/10.7324/JAPS.2025.260740

Li, H., Yang, Y., Hong, W., Huang, M., Wu, M., & Zhao, X. (2020). Applications of genome editing technology in the targeted therapy of human diseases: Mechanisms, advances and prospects. *Signal Transduction and Targeted Therapy, 5*(1), 1. https://doi.org/10.1038/s41392-019-0089-y

Loewen, G., & Pollard, W. (2010). The social justice perspective. *Journal of Postsecondary Education and Disability, 23*(1), 5–18.

Martz, E. (2001). Other research—Acceptance of imperfection. *Disability Studies Quarterly, 21*(3). https://doi.org/10.18061/dsq.v21i3.302

Matthewson, J., & Griffiths, P. E. (2017). Biological criteria of disease: Four ways of going wrong. *Journal of Medicine and Philosophy, 42*(4), 447–466. https://doi.org/10.1093/jmp/jhx004

McKenzie, A. T., Wowk, B., Arkhipov, A., Wróbel, B., Cheng, N., & Kendziorra, E. F. (2024). Biostasis: A roadmap for research in preservation and potential revival of humans. *Brain sciences, 14*(9), 942. https://doi.org/10.3390/brainsci14090942

McKinney, C. (2021). Biopluralism, disability, and democratic politics. *Politics, Groups, and Identities, 9*(2), 423–437. https://doi.org/10.1080/21565503.2021.1877750

Mendz, G. L., & Cook, M. (2021). Posthumanism: Creation of 'new men' through technological innovation. *The New Bioethics, 27*(3), 197–218. https://doi.org/10.1080/20502877.2021.1953266

Merzlyakov, S. S. (2022). Posthumanism vs. transhumanism: From the "end of exceptionalism" to "technological humanism". *Herald of the Russian Academy of Sciences, 92*(Suppl. 6), S475–S482.

Mirazón Lahr, M. (2016). The shaping of human diversity: Filters, boundaries and transitions. *Philosophical Transactions of the Royal Society B: Biological Sciences, 371*(1698), 20150241. https://doi.org/10.1098/rstb.2015.0241

Moeller, A., & Andres Porras, J. (2025). Human enhancement, past and present. *Monash Bioethics Review*. https://doi.org/10.1007/s40592-025-00250-5

Ne'Eman, A., & Bascom, J. (2020). Autistic self advocacy in the developmental disability movement. *The American Journal of Bioethics, 20*(4), 25–27. https://doi.org/10.1080/15265161.2020.1730507

Oliveira, D., Martins, C., Faísca, L., Brás, M., Nunes, C., & Carmo, C. (2025). The role of parental perfectionism and child temperament in the intergenerational transmission of perfectionism: A pilot study. *Children, 12*(11), 1452. https://doi.org/10.3390/children12111452

Palazzani, L. (2020). Reproductive technologies and the global bioethics debate: A philosophical analysis of the report on ART and Parenthood of the International Bioethics Committee of Unesco. *Phenomenology and Mind, 19*, 138–149. https://doi.org/10.17454/pam-1910

Pourkhalili, S., Shal, R. S., Abolghasemi, A., Dianatkhah, M., & Gharipour, M. (2022). *The share of genetic and environmental factors to perfectionism: A classical twin study*. https://doi.org/10.21203/rs.3.rs-2196363/v1

Pugh, K. C. (2013). Genetic engineering and the pursuit of human perfection. *CedarEthics: A Journal of Critical Thinking in Bioethics, 13*(1), 6. https://doi.org/10.15385/jce.2013.13.1.6

Roduit, J. A., Baumann, H., & Heilinger, J. C. (2013). Human enhancement and perfection. *Journal of Medical Ethics, 39*(10), 647–650. https://doi.org/10.1136/medethics-2012-100920

Romanska, M. (2024). The bionic body: Technology, disability and posthumanism. *Body, Space & Technology, 23*(1), 1–17. https://doi.org/10.16995/bst.11480

Samuel, R., & Jacob, K. S. (2018). Empowering people with disabilities. *Indian Journal of Psychological Medicine, 40*(4), 381–384. https://doi.org/10.4103/IJPSYM.IJPSYM_90_18

Sandel, M. (2007). *The case against perfection: Ethics in the age of genetic engineering*. Belknap Press of Harvard University Press. https://doi.org/10.1007/s12376-009-0018-4

Sauer, J. S., & Lalvani, P. (2017). From advocacy to activism: Families, communities, and collective change. *Journal of Policy and Practice in Intellectual Disabilities, 14*(1), 51–58. https://doi.org/10.1111/jppi.12219

Saur, E., & Sidorkin, A. M. (2018). Disability, dialogue, and the posthuman. *Studies in Philosophy and Education, 37*, 567–578. https://doi.org/10.1007/s11217-018-9616-5

Sen, A. (2000). Development as freedom. *Development in Practice—Oxford, 10*(2), 258–258.

Siham, M. A., Prabahar, A. A., Nixon, S., Sandhiya, P., Sowndariyan, G. G., & Shagar, V. S. (2023). Investigating the frontiers of genetic regulation and unlocking the secrets of gene expression: The power of RNA-based therapeutics. *Journal of Clinical Genetics and Heredity, 1*(1), 19–41. https://doi.org/10.37191/Mapsci-JCGH-1(1)-002

Slaney, R. B., Rice, K. G., Mobley, M., Trippi, J., & Ashby, J. S. (2001). The revised almost perfect scale. *Measurement and Evaluation in Counseling and Development, 34*, 130–145. https://doi.org/10.1080/07481756.2002.12069030

Suarez-Balcazar, Y., Balcazar, F., Labbe, D., McDonald, K. E., Keys, C., Taylor-Ritzler, T., et al. (2023). Disability rights and empowerment: Reflections on AJCP research and a call to action. *American Journal of Community Psychology, 72*(3–4), 317–327. https://doi.org/10.1002/ajcp.12710

Tamura, R., & Toda, M. (2020). Historic overview of genetic engineering technologies for human gene therapy. *Neurologia Medico-Chirurgica, 60*(10), 483–491. https://doi.org/10.2176/nmc.ra.2020-0049

Triviño, J. L. P. (2015). Equality of access to enhancement technology in a posthumanist society. *Dilemata, 19*, 53–63.

United Nations. (2017). *Development*. Convention on the Rights of Persons with Disabilities (CRPD). Retrieved from https://www.un.org/development/desa/disabilities/convention-on-the-rights-ofpersons-with-disabilities.html

Van Niekerk, A. A. (2014). Biomedical enhancement and the pursuit of mastery and perfection: A critique of the views of Michael Sandel. *South African Journal of Philosophy, 33*(2), 155–165. https://doi.org/10.1080/02580136.2014.923697

Veenhoven, R. (2010). Life is getting better: Societal evolution and fit with human nature. *Social Indicators Research, 97*(1), 105–122. https://doi.org/10.1007/s11205-009-9556-0

Veit, W. (2021). Biological normativity: A new hope for naturalism? *Medicine, Health Care and Philosophy, 24*(2), 291–301. https://doi.org/10.1007/s11019-020-09993-w

Vieira, D. K. R., & Favoreto, C. A. O. (2016). Narratives on health: Reflections on care for people with disabilities and genetic disease within the Brazilian National Health System (SUS). *Interface-Comunicação, Saúde, Educação, 20*, 89–98. https://doi.org/10.1590/1807-57622015.0203

Wall, S. (2007). *Perfectionism in moral and political philosophy*.

Youvan, D. C. (2024). *The ethics of genetic enhancement: Examining the societal consequences of unregulated gene editing for the elite*.

Chapter 4
Euthanasia, End-of-Life Ethics, and Disability Bias

Abstract This chapter investigates euthanasia and assisted dying from the interconnected perspectives of vulnerability, disability bias, and healthcare distributive justice. People with disabilities are often depicted as being vulnerable by nature, a perception that is supported by medicalized and deficit-based models that undermine autonomy and strengthen paternalism. This chapter relies on relational interpretations of vulnerability and the CRPD framework, stressing that vulnerability comes from social, structural, and environmental interactions, not from a person's impairment. The chapter provides a thorough examination of medical futility, DNRs, and end-of-life choices, and highlights difficulties faced by disabled patients, including differing interpretations, cultural differences, and disagreements over power levels. The debates about resource distribution, such as triage criteria and ventilator shortages during the COVID-19 pandemic, and the ethical limits of QALYs, structural inequities that have always been faced by disabled people, particularly in times of crisis, are discussed. In general, the chapter brings together legal, ethical, and clinical aspects to demonstrate how end-of-life practices can either perpetuate or eliminate discrimination and calls for the establishment of safeguards that will support the principles of autonomy, equity, and inclusive justice.

Keywords Vulnerability · Ableism · Disability bias · Diagnostic overshadowing · Resource allocation

4.1 Introduction

Euthanasia ethics have been debated since ancient Greece and Rome. Euthanasia is a sensitive topic, and the majority of the discussion around it has taken place in referendum slogans and the media. As a result, there has been a typical disregard for significant differences and arguments, and the inability to discern between regions of agreement and disagreement (Emanuel, 1994). The Greek terms "*eu*" and "*thanatos*" form the basis of the word "euthanasia," which means "good death." When

A. Shibi Anilkumar, R. Veerabathiran, *Bioethics and Disability*, SpringerBriefs in Modern Perspectives on Disability Research,
https://doi.org/10.1007/978-981-95-8197-9_4

death cannot be avoided, euthanasia may be the choice for some people to relieve their suffering, maintain their honor, or shorten their dying (Gupta & Bansal, 2023). It is difficult to tell the difference between euthanasia and physician-assisted suicide (PAS) since both cases imply the desire to live no more and, therefore, the acceptance of assisted death. To separate one from the other, it has to be clarified what the last step is that leads to the victim's death. In death by euthanasia, the clinician gives the injection, while the forbearer takes the medication, which they receive from the doctor, until overdose is reached and their death occurs in PAS (Gupta & Bansal, 2023).

Clinical, legal, political, religious, and ethical factors all play a significant part in the ongoing moral controversy surrounding assisted suicide. There has been discussion about asking for assisted suicide and refusing or stopping treatment in connection with the concurrent global expansion of palliative care (Fontalis et al., 2018). Since they come under the category of life as a human right, which has been widely upheld for several years, euthanasia and assisted suicide have been considered all over history. Many historical occurrences, including the Nazi experiments, associated "euthanasia" more with killing than with an act of kindness and compassion. According to more recent writings, euthanasia is the process of hastening the demise of a patient with an untreatable or deadly illness by using or refraining from clinical treatments to prevent undue suffering (Picón-Jaimes et al., 2022).

Only a few nations or areas of the world now allow euthanasia. Euthanasia can be classified into three major categories: active, passive, and indirect. Indirect euthanasia is the performance of an act that is intended to relieve or stop the suffering, even though it might temporarily shorten the life, while passive euthanasia is allowing the patient to decease by holding back or ceasing life-supporting intervention, etc. In countries where euthanasia is permitted, although the conditions for euthanasia have been laid out, several problems have arisen with the practice itself. Numerous academic fields, including medicine, psychology, pharmacology, and ethics, are involved in end-of-life research (Kono et al., 2023).

The patient's well-being, being the primary consideration, is an ethical imperative that underlies medical practice. It provides a set of rules for conduct and decision-making that enable physicians to understand the expectations of their patients, professional colleagues, and society as a whole. Additionally, the World Medical Association (WMA)'s moral guidelines serve as the basis for determining which behaviors on the part of doctors are appropriate in these matters. One of three strategies typically supported by legislators is a complete ban on euthanasia, comparing it to regular or privileged murder, or allowing it under specific guidelines (Mousavi, 2024).

The ethical perspective of not taking another person's life should not be unqualified. There should be exceptions where granting a patient's request to be assisted in dying would be morally acceptable and likely required. Specifically, certain situations should be found where, as an extreme ratio, only assisted suicide is a legitimate requirement to protect an individual's fundamental rights, such as the following: the right to health, which, despite being irreversibly destroyed by the illness, remains the privileges not to continue in an unbearable condition of suffering; and the right to self-respect, self-governance, and safeguarding moral identity (Scopetti et al., 2023).

The most influential argument against the authorization of assisted dying is the appeal to the rights of people with disabilities (PwDs). Various disability rights groups and people with disabilities express different views on the subject of permitting assisted dying. To generalize is to be disrespectful and misinterpret the position of the disabled community, who have varying opinions. Laws regarding assisted suicide are degrading to people with disabilities, particularly when they carry the message that their lives are not meaningful or at least not as worthy as those of others. It is often asserted that the laws governing assisted death lead to the situation where PwDs are "threatened, manipulated, or forced" to choose assisted dying. This argument centers on the substantive issue of what respect for disabled lives and autonomy necessitates (Colburn, 2022).

The eugenics movement, which took place mainly in the late nineteenth and early to mid-twentieth centuries, was focused on the control and, in several instances, the elimination of people considered "unfit." The Nazi Aktion T4 program, which was a precursor and testing ground for the overall Holocaust, represented the highest point of the eugenics movement. About 300,000 persons with disabilities were murdered by doctors and the state as a result of Aktion T4, many of them with the intention of putting an end to their suffering. As more and more jurisdictions think about permitting or expanding euthanasia and assisted dying (EAS), we must consider the very real concerns of the disability community and make sure that, in our desire to increase the autonomy of the many, we do not infringe upon the rights of disabled people to live a life of quality and equity (Stainton, 2022). Thus, this chapter examines euthanasia and assisted dying practices through the lens of vulnerability, bias, justice, and resource allocation.

4.2 Vulnerability

Some groupings of PwDs, in particular, are often recognized as vulnerable. Such charges that states owe toward PwDs, as "vulnerable," have been emphasized in several cases of international human rights law. The language of vulnerability and disability, and the stigmatization and stereotyping that such ideas can be utterly marginalizing and dehumanizing. This is true mainly for PwDs, given that impairments in many situations remain to be addressed in bad terms of fixing a "weakness" or a "deficit," along the medical approach to disability, seeing disability as a consequence of a person's impairment. In many ways, PwDs are hence stuck in perceptions of vulnerability shaped by how society responds to different exposures of reliance and autonomy (Mustaniemi-Laakso et al., 2023).

4.2.1 *Universal Versus Group-Based Vulnerability*

Fineman opts for a reformed understanding of what it means to be human (vulnerability and dependence instead of rationality and autonomy) and an understanding of vulnerability that underlines its general character; others, however, contest the vulnerable-invulnerable binary by noting that vulnerability is essentially a matter of degree (Fineman, 2012). In disability studies, the notion of "vulnerability" has been controversial. It has been linked to the medical paradigm of disability, which presents disability as a personal tragedy or an individual issue (Burghardt, 2013). According to academics from various fields, "social vulnerability" is a pre-existing state of social structures in which certain social factors, such as cramped living conditions, worsen the effects of natural disasters on marginalized populations (Bergstrand et al., 2015).

4.2.2 *Human Vulnerability in International Bioethics*

According to UNESCO's International Bioethics Committee report from 2013, "The Principle of Respect for Human Vulnerability and Personal Integrity," states vulnerability can be viewed as both a concept that is particularly pertinent to a select few and something that every human being is exposed to as part of the "Human Condition" (International Bioethics Committee, 2013). Minorities, the elderly, those with illnesses that drastically reduce life expectancy, those with mental illnesses, substance abusers, children, and those with intellectual disabilities are among individuals who are seen to require particular protection. In addition to the risk that comes with simply being human, vulnerable populations are frequently defined as individuals who are thought to have a higher chance of experiencing harm, abuse, or other unfavorable outcomes in life and in study. "Some communities and individuals are predominantly weak and may have an increased likelihood of being maltreated or of experiencing extra harm," according to the Helsinki Declaration. Every vulnerable group and individual should be given special consideration for protection (Snipstad, 2022).

4.2.3 *Structural and Contextual Vulnerabilities*

Since over 80% of disabled people reside in countries with limited resources, many of them are incredibly vulnerable to the effects of climate change and natural disasters (Engelman et al., 2022). A person's vulnerability to human trafficking may be increased if they have a disability, according to several national and international reports. Due to their social and economic conditions, as well as the shame and marginalization they experience in their communities, people with disabilities may be motivated to look for opportunities in exploitative situations. Unfair power relations may make people more vulnerable to caregiver exploitation (Jagoe et al., 2025).

4.2.4 *Relational Models of Vulnerability and Disability*

Due to our physical dependence on and need for others, vulnerability becomes a necessary condition for humans; some kinds are constant, whereas others vary with age, health, or disability. By separating these sources, vulnerability is seen to result from interactions between people and their surroundings, with situational vulnerability influenced by environmental, social, economic, or personal factors. This aligns with the UN Convention on the Rights of Persons with Disabilities' characterization of the concept of disability as a dynamic, relational term coming from the interaction of environmental or attitudinal barriers and impairments. As a result, disability develops in settings influenced by relationships, politics, culture, and history. This viewpoint represents a shift from a medical paradigm that emphasizes rehabilitation and correction to a social and human rights-based approach that seeks to eliminate institutional barriers to equal participation and views disability as a universal human experience (Mustaniemi-Laakso et al., 2023).

4.2.5 *Autonomy, Paternalism, and Ethical Tensions*

In an effort to place vulnerability within conceptions of all people as interdependent and relational, and thus ontologically universal, many theorists are challenging conventional approaches to vulnerability, which saw it as a group- or status-based concept tied to an innate condition such as age or disability. Vulnerability as a social policy indicator can be risky, as it may lead to actions that normalize "acceptable" behaviors, perpetuate power dynamics, and hinder the development of resilience or personal preventive strategies. This perspective views interference in someone's decision-making as paternalistic and disrespectful of one's autonomy. Thus, the only time one should argue against a person's choice is when the situation is particular, to avoid unjustified and unethical coercion of individual freedom. Thus, the conflict between paternalism and autonomy is evident (Clough, 2017).

4.2.6 *Power Imbalance and Coercion*

Power imbalance was also mentioned as a state linked to vulnerability. If you depend on someone to provide a specific need for you, you may perceive them as having power. A person with an intellectual disability is disempowered when there is a power imbalance. In addition to fostering a culture of power imbalance, when someone depends too much on others or systems for help, those who provide the help gain more control. This makes the dependent person more vulnerable and increases the risk of mistreatment or exploitation (Parley, 2011).

4.2.7 Decision-Making Capacity and Autonomy Risks

The task-specific ability to comprehend and value pertinent information, weigh options, and explain a decision is known as decision-making capability. People with cognitive disabilities range in their ability to make decisions. People are competent at one extreme and totally disabled at the other. A person with limited capacity is susceptible to both autonomy violations and her own possibly damaging decisions (Peterson et al., 2021; Akila et al., 2020).

4.3 Disability Justice Critiques of Assisted Suicide

4.3.1 Ableism and Structural Inequities in Healthcare

A way of thinking and acting that favors some bodies and minds over others is known as ableism. This type of discrimination centers on the idea of "normal" within a value hierarchy in which "abled" is superior to "disabled". Ableism is a systemic healthcare issue, as evidenced by the shocking fact that more than 80% of doctors assume that patients with impairments have a worse condition of life than those without disabilities. Change will be facilitated through the implementation of suggested anti-ableism initiatives by collaborating with the current generation of practicing clinicians and staff champions dedicated to enhancing the integration of disability in medical education. It is crucial to understand that this article and these concepts are by no means the first call to action for standardized curriculum competences on disability education in healthcare (Kaundinya & Schroth, 2022).

4.3.2 Disability Justice Movement and Rights-Based Activism

Following the disability civil rights movement, a worldwide disability justice movement emerged in 2005. The civil rights movement focused on rights-based activism and single social identity-based disability efforts, despite its many achievements (Goulden et al., 2023).

Outside the disability studies field, there hasn't been much theoretical work on the denial of a right to PAS. The denial of a right has significant ramifications for disability activism. The bases behind the disability rights movement's denial of that right to independence and self-determination should be examined by disability studies scholars, who should also consider the consequences for how disability activism presents itself in relation to other progressive groups. In turn, sociolegal studies can use PAS's disability denial as a helpful framework for analyzing the function of rights in social movements (Heyer, 2011).

4.3.3 *Social Devaluation, Healthcare Inequities, and Resource Obligations*

Recognizing the prevalence of social devaluation based on class, age, and disability, as well as the injustices of our nation's healthcare system, including the increasing physical risk that many patients face in hospitals due to cost-cutting measures and shifting standards of care, it is necessary to acknowledge the validity of the disability rights movement's opposition to assisted suicide. Advocates must confront their own sentiments toward resource sharing and social obligations to concede such points (Gill, 2010).

They have to face their own possible vulnerability in the event of an incurable illness, loneliness, or poverty, as well as deeply ingrained fears about being disabled. Supporters who genuinely comprehend disability rights reasons against assisted suicide may have to give up cherished identities as advocates of autonomy on top of all those unsettling revelations and related hazards to peace of mind (Gill, 2010).

4.3.4 *Social Perceptions of Disability and the Risk of Devaluing Disabled Lives*

The majority's belief that a life with a disability is inherently less valued than a life without one may be reflected in laws like PAS. Even though this viewpoint is harmful in that it promotes a hierarchy of human worth that has nothing to do with an individual's self-evaluation, its actual danger lies in its purportedly likely extension into the idea of "better dead than disabled," if it is indeed prevalent in the healthcare industry as well as elsewhere. As a result, in cases of severe disability, procedures like euthanasia are clearly acceptable. Unbearable and inevitable current or impending pain is typically thought to be the driving force for both PAS and euthanasia; nevertheless, the somewhat ambiguous quality of suffering is believed to cover those with disabilities more extensively (Shildrick, 2008).

4.3.5 Discriminatory Framing and Paternalistic Overemphasis on Vulnerability

"Discriminatory against people with disabilities" was the stance. Scholars who oppose assisted dying laws because they treat PwDs "as some 'unknown', 'disabled person' lacking a personality" and "as unskilled, easily bullied, and inclined to end their lives, points them in the roles to which they have been limited by disability discernment" also express this concern in academic literature. Many disabled persons "complain that it feeds rather than starves social preconceptions, finding this stereotyping to be itself demeaning and patronizing." The risk of a "authoritarian over-focus on the vulnerability of persons with disabilities" is acknowledged even by some opponents of assisted suicide (Colburn, 2022).

4.3.6 Legislative Developments and the Slippery Slope Concerns

On November 29, 2024, the House of Commons gave the green light to the Terminally Ill Adults (End-of-Life) Bill, which is a Private Members' Bill put forward by Kim Leadbetter. The objective of the Bill is to "Permit adults who are fatally ill, subject to defenses and protections, to demand and be provided with aid to end their own life." According to the Bill's followers, strong precautions will be put in place to inhibit the extension of the criteria, as has occurred in other jurisdictions. Opponents of the Bill are frequently alleged of slippery-slope arguments, which are easily dismissed when they express matters that the proposed safeguards are likely to be unproductive and that, should the regulation be passed, the scope of practice will certainly expand, endangering the vulnerable. A slippery slope is a metaphorical slope where the position that is most disagreeable from the top is at the bottom, and the most meritorious position is at the top (Trimble, 2025).

The slippery slope argument is presented differently. It argues that if reasonably foreseeable is eliminated, there is no conceptual justification for restricting assisted suicide to individuals whose suffering is deemed intolerable because of, say, a chronic illness. The slippery slope issue here is this: logically, what would prevent us from offering assisted dying to someone who might end up in extreme poverty and choose to request an assisted death? Because of their poverty, they may believe that their suffering is unbearable (Trimble, 2025).

4.4 Medical Futility, DNR Orders, and Disability Bias

4.4.1 Medical Futility

According to the American Medical Association, futility cannot be meaningfully defined and is described as "inadequacy to produce a result or bring about a required end; ineffectiveness" (Rolak et al., 2024). The "goals of medicine" paradigm distinguishes two interconnected meanings: quantitative futility, referring to physiologic inefficacy when an intervention has an extremely low chance of achieving the intended physiologic outcome; and qualitative futility, evaluating whether an intervention accomplishes a legitimate medical goal and provides meaningful benefit to the patient. Because qualitative futility relies on normative judgments, there is significant ethical and legal disagreement about whether clinicians should have this authority (Kopar et al., 2022).

4.4.2 Ethical Disputes Over Futility

Disputes concerning futility very often originate in fundamentally different existential frameworks rather than mere information gaps. The patients and the doctors are on different meaning-making paths; therefore, futility judgments should be considered as continuous interpretative dialogues, not rigid clinical criteria (Huang, 2025). Technological advances can prolong the lives of terminally ill patients even when underlying illnesses are untreatable, raising concerns about ICU resource shortages and healthcare costs, especially among older, chronically ill patients (Aghabarary & Dehghan Nayeri, 2016). Conflicts also arise between physician authority and patient autonomy, with some clinicians asserting the primacy of medical judgment, while families may insist on continuing treatment even when clinicians deem it ineffective (Meltzer & Huckabay, 2004; Saettele & Kras, 2013).

4.4.3 Do-Not-Resuscitate (DNR) Orders and End-of-Life Practices

Some patients' conditions deteriorate to the point where physicians issue do-not-resuscitate (DNR) orders, which must be implemented by critical care nurses (Ntseke et al., 2023). Cardiopulmonary resuscitation (CPR) in cardiac arrest involves a range of basic and advanced interventions, but may lead to poor outcomes; long-term survival after CPR is often low (Pachys et al., 2014).

Do-not-resuscitate (DNR) orders do not restrict routine care; patients continue receiving antibiotics, pain relief, and other treatments. Introduced about 50 years ago, DNR guidelines emerged due to CPR's frequent ineffectiveness and high

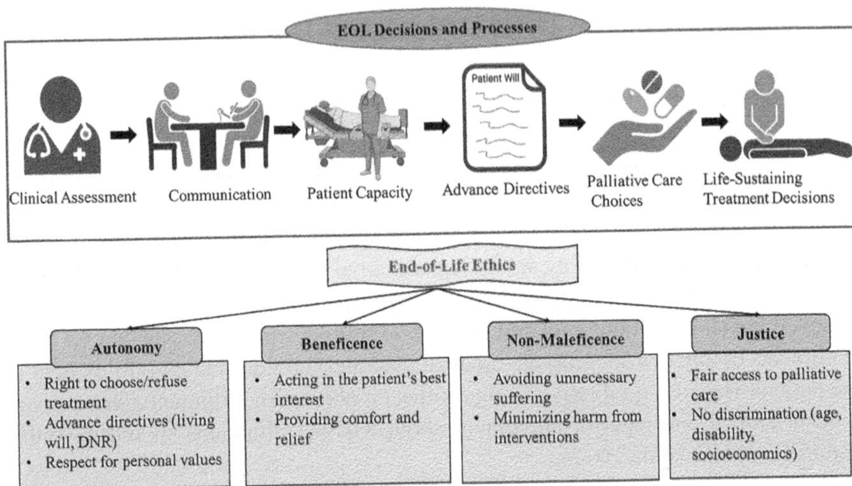

Fig. 4.1 Ethical principles and decision-making processes at the end-of-life. The figure presents a visual overview of end-of-life (EOL) decisions, processes, and ethical principles. The top section, labeled "EOL Decisions and Processes," includes icons representing a physician consultation, shared decision-making between patient and provider, a hospitalized patient under medical supervision, a patient will document, and a comparison of medication choices with resuscitation (CPR). The lower section, titled "End-of-Life Ethics," outlines four core ethical principles. Autonomy includes the right to choose or refuse treatment, to use advance directives, and to uphold personal values. Beneficence focuses on acting in the patient's best interest and providing comfort. Nonmaleficence emphasizes avoiding unnecessary suffering and minimizing harmful interventions. Justice highlights fair access to palliative care and the absence of discrimination based on age, disability, or socioeconomic factors

financial and human costs (Raoofi et al., 2021). Do-not-resuscitate (DNR) decisions are typically made for terminally ill patients receiving palliative care and may or may not involve patient input, especially when cognitive impairment or severe illness limits decision-making capacity (Paratz et al., 2023). Figure 4.1 summarizes key components of end-of-life decision-making and the ethical principles that guide equitable end-of-life care.

4.4.4 *Legal Authority and Surrogate Decision-Making*

Courts vary in how they treat end-of-life decisions for incompetent patients. There are different opinions regarding the guardianship power. Some are of the view that guardians should possess comprehensive power to make end-of-life decisions to protect the patient's best interests, while others think that disabled patients are better protected when guardian power is limited (Moreno, 1995). Conflicts of interest may occur: the guardians may deny treatments to lessen the burden of caregiving, or they may keep the patient alive for financial reasons related to the guardianship payment.

To reduce these risks, it is recommended that court approval be obtained for withholding or withdrawing life-saving treatment (Leever et al., 2012).

4.4.5 Undesired CPR Events

Factors contributing to undesired CPR include physicians overriding code status, failure to disseminate and honor advance directives, and unawareness of code status. Undesired CPR also occurs when physicians view the arrest as caused by physician error or procedural complications. Frequent hand-offs and unfamiliarity with the patient by resuscitation teams increase the likelihood of performing CPR despite DNR orders (Wong et al., 2020).

4.4.6 Cultural and Institutional Variability in DNR Policies

Hospitals differ in their DNR policies based on institutional priorities and cultural norms. For example, DNR orders are not accepted in Japan. In Saudi Arabia, Islamic law guides decisions: since 1988, DNR is allowed only when three knowledgeable physicians agree the condition is hopeless, excluding family input. In 2004, the NIH recognized end-of-life care as essential to care in the final stages of life (Alsaati et al., 2020).

4.4.7 Disability Bias and Barriers in End-of-Life Care

Implicit and explicit biases among clinicians toward disabled patients remain understudied. Even experienced physicians may fail to actively include disabled patients in care, and many disabled individuals report that clinicians underestimate their quality of life. This contrasts with the "disability paradox," that over half of disabled people describe their quality of life as excellent or good despite public assumptions to the contrary (Iezzoni et al., 2021). Intensive care unit (ICU) admission disparities also occur among older adults, women, racially diverse groups, and persons with long-term neurological disabilities who may be denied critical care due to perceived lower quality of life. The majority of ICU doctors and nurses are not trained in the specific issues of disability (Taccone, 2023). Individuals with intellectual disabilities encounter enormous organizational, social, and physical barriers when trying to get healthcare of the right quality and in time, although their physical health problems are very high. Hospital admission can be particularly challenging due to systemic obstacles (Dunn et al., 2018).

4.5 Clinical Decision-Making and Implicit Bias

Implicit biases are mental associations that affect a person's interactions with others but are not conscious or under their control. Implicit biases cause people to behave in ways that are inconsistent with their explicit principles or beliefs. Behaviors in socially delicate situations, such as interracial interactions, are linked to unconscious biases (Salles et al., 2019). The term *bias* raises major dilemmas in the healthcare sector and is often applied to implicit stereotypes and prejudices. The presence of implicit biases among healthcare professionals and their impact on the level of medical treatment should be of concern. Healthcare providers are concerned about unconscious biases that work against people who are already at risk (FitzGerald & Hurst, 2017). Implicit bias and structural racism are mutually reinforcing; an entity's cognitive links between cultural identity and worse consequences (implicit bias) are reinforced by ambient structural racism and its consequences. Stereotypes of marginalized groups or communities, as well as tacit and explicit prejudices against them, have been produced by structural injustices and discrimination. "Preferences, principles, and attitudes of which people are generally consciously aware, endorsed, and can be acknowledged and communicated" are examples of explicit kinds of prejudice (Vela et al., 2022).

"Diagnostic overshadowing" refers to a "adverse bias affecting a clinician's decision regarding co-occurring illnesses in individuals who have learning disabilities or other mental illnesses." Although this phenomenon is not frequently included in lists of cognitive and clinical biases, awareness of it as a kind of prejudice in healthcare is still growing (Hallyburton, 2022). It is becoming more widely acknowledged that diagnostic overshadowing contributes to the health disparities that people with learning disabilities confront (Javaid et al., 2019). Time is a necessary resource for both emotional and cognitive deliberation, but if it is overspent, opportunities may be lost, and no practical conclusion can be reached. When under time constraint, people tend to make decisions more quickly, use more heuristics, produce lower-quality results, and make biased decisions. However, there hasn't been enough research on emergency decision-making involving human life, which is one of the most common situations in which time is of the essence (Taheri et al., 2023).

Research indicates that educating healthcare professionals in cultural competence, including navigating the culture of disability, can enhance their knowledge, attitudes, and abilities. The ability of providers to successfully involve patients in their own care is also positively affected by training intended to encourage patient-centered care that emphasizes the whole person rather than a single component, such as a specific ailment, diagnosis, or disability (Phillips et al., 2021).

The detrimental effects of stigma against people with disabilities are effectively mitigated by providing employers with specialized training that addresses implicit and explicit bias. People with disabilities are more prevalent in the workforces of many organizations that have implemented explicit disability inclusion policies and procedures (Bezyak et al., 2024).

4.6 Resource Allocation Ethics

Need (giving resources exclusively to those who cannot survive without them) and benefit (providing resources to those most likely to survive with their use) are examples of triage criteria. However, there is significant disagreement over the most effective way to evaluate need and benefit, as well as whether any other criteria should be included (Antommaria et al., 2020).

Large cohorts of patients with specific concurrent conditions, such as class III or IV heart failure, severe chronic lung disease, end-stage kidney disease, and severe cognitive loss, are prohibited from entering intensive care units (ICUs) by professional society guidelines and some state recommendations as the coronavirus disease 2019 (COVID-19) epidemic worsens. Because the exclusion criteria (long-term clinical outcome and functional status) are applied exclusively to certain patient types rather than to all patients examined for critical care, these exclusions are unethical and lack explicit justification (White & Lo, 2020).

The COVID-19 pandemic has caused an unexpected problem for hospitals and medical professionals: a shortage of ventilators, with the dilemma of taking ventilators off certain patients so others with a higher chance of survival can get them. When mechanical ventilation is no longer aligned with a patient's objectives or instructions, care teams typically do not initiate or maintain it. COVID-19 creates a contrasting scenario: many patients who were refused a ventilator may have lived for many more years if they had been given short-term ventilation. Refusing this treatment would free up a limited resource for patients who have a higher chance of survival with ventilatory assistance, even if it would go against their desires and probably result in their death (Cohen et al., 2020).

Medical ethics has long discussed how to distribute limited resources, and procedural standards have been established. The prioritization process must be clear, inclusive (permitting contribution by all individuals who are impacted by decisions emerging from the process), evidence-based, and adaptable to new information or arguments to claim moral legitimacy (Joebges & Biller-Andorno, 2020).

Quality-Adjusted Life Years (QALYs) were created to facilitate the efficient allocation of healthcare resources by identifying the health outcomes that can be produced most effectively (i.e., at the lowest cost) and consequently maximize population health. By offering a standard metric for openly and quantitatively assessing the effects of various illnesses and treatments on health, QALYs facilitate system-wide decision-making (Rand & Kesselheim, 2021).

While QALYs are the recommended outcome metric for performing cost-effectiveness analysis (CEA) in wealthy nations, DALYs are preferable in underdeveloped countries (Deshmukh et al., 2013). Prolonging the lives of PwDs or underlying medical conditions results in fewer QALYs than extending the lives of "healthier" people, according to a long-standing criticism that the QALY differentiates against those with disabilities and those in poor health (Schneider, 2022).

People with impairments are still denied equality, typically in the form of laws that limit legal competence, such as plenary or partial guardianship. The CRPD established equal recognition before the law under Article 12 and marked a shift from substituted decision-making to supported laws and procedures for decision-making (Holness, 2014).

However, given the extreme scarcity that low-income nations face, it might not be feasible to distribute regionally accessible health resources sufficiently to enable citizens to attain a satisfactory outcome of health. Because there are inadequate resources to offer portions sufficient for everybody to achieve acceptable levels of health, specific allocations are destined to be unfair. Therefore, when allotting healthcare resources and modifying institutional arrangements that impact people's health, besides relying on the typical normative principles of fairness, priority, and productivity, the question of thresholds of sufficiency of resources required to attain and preserve marginally acceptable health arises (Alvarez, 2007).

The randomization principle has received support, particularly from organizations that promote the rights of those with disabilities. Better, quicker access to healthcare may be available to patients who are socially better off, potentially leading to the selection of wealthy individuals and possibly discrimination against less fortunate patients who arrive at hospitals later. This may also impact people with disabilities, as they may have equal treatment opportunities but not always equal access to healthcare. Social injustice and adversity are issues that people with disabilities have to deal with quite often. Consequently, a first-come, first-served approach will likely increase health inequalities among people with disabilities (Göttert, 2025). Denial of any discrimination is a prerequisite for the realization of inclusive equality in triage (Degener, 2024).

4.7 Conclusion

It is therefore difficult to apply ethical principles solely to end-of-life situations without considering the wide-ranging social, political, and structural factors that shape perceptions and the management of disability in healthcare. The chapter illustrates that notions such as vulnerability, medical futility, and resource allocation are not neutral but are influenced by social devaluation, implicit bias, and ableism with roots in history. Disability justice critiques highlight the potential unconscious support of stereotypes that regard the lives of the disabled as a burden or less valuable, especially when paternalistic assumptions in the context of assisted dying overshadow autonomy. The same applies to DNR orders, surrogate decision-making, and judgments about futility, as they might cause disabled people to be more easily coerced, excluded, or their quality of life misunderstood. Along with the aforementioned factors, structural barriers, improper training, and diagnostic overshadowing also make the situation worse. Controversies over the allocation of resources, such as ventilator triage during the COVID-19 pandemic, and the use of QALYs show that crisis ethics can worsen the situation for the already marginalized unless they are designed with inclusivity, transparency, and fairness as their core values. Human rights-based safeguards, particularly those outlined in the CRPD, are indispensable for protecting autonomy, dignity, and equal recognition under the law. Thus, ethical decision-making in dying situations requires not only the active dismantling of prejudice, enhancement of safeguards, and ensuring that the disabled have a say equal to that of others in matters of their own life.

References

Aghabarary, M., & Dehghan Nayeri, N. (2016). Medical futility and its challenges: A review study. *Journal of Medical Ethics and History of Medicine, 9*, 11.

Akila, K., Divya, R., Preethianushya, M., Aravindhan, B., & Rogina, J. S. (2020). Assessment of cognitive impairment among elderly in the selected rural community, Kancheepuram District, Tamil Nadu. *Indian Journal of Public Health Research & Development, 11*(3), 19–21. https://doi.org/10.37506/ijphrd.v11i3.599

Alsaati, B. A., Aljishi, M. N., Alshamakh, S. S., Basharaheel, H. A., Banjar, N. S., Alamri, R. S., & Alkhayyat, S. (2020). The concept of do not resuscitate for the families of the patients at king Abdul-Aziz University hospital. *Indian Journal of Palliative Care, 26*(4), 518. https://doi.org/10.4103/IJPC.IJPC_228_19

Alvarez, A. A. A. (2007). Threshold considerations in fair allocation of health resources: Justice beyond scarcity. *Bioethics, 21*(8), 426–438. https://doi.org/10.1111/j.1467-8519.2007.00580.x

Antommaria, A. H. M., Gibb, T. S., McGuire, A. L., Wolpe, P. R., Wynia, M. K., Applewhite, M. K., et al. (2020). Ventilator triage policies during the COVID-19 pandemic at US hospitals associated with members of the association of bioethics program directors. *Annals of Internal Medicine, 173*(3), 188–194. https://doi.org/10.7326/M20-1738

Bergstrand, K., Mayer, B., Brumback, B., & Zhang, Y. (2015). Assessing the relationship between social vulnerability and community resilience to hazards. *Social Indicators Research, 122*(2), 391–409. https://doi.org/10.1007/s11205-014-0698-3

Bezyak, J., Versen, E., Chan, F., Lee, D., Wu, J. R., Iwanaga, K., Rumrill, P., Chen, X., & Ho, H. (2024). Needs of human resource professionals in implicit bias and disability inclusion training: A focus group study. *Journal of Vocational Rehabilitation, 60*(3), 311–319.

Burghardt, M. (2013). Common frailty, constructed oppression: Tensions and debates on the subject of vulnerability. *Disability & Society, 28*(4), 556–568. https://doi.org/10.1080/09687599.2012.711244

Clough, B. (2017). Disability and vulnerability: Challenging the capacity/incapacity binary. *Social Policy and Society, 16*(3), 469–481.

Cohen, I. G., Crespo, A. M., & White, D. B. (2020). Potential legal liability for withdrawing or withholding ventilators during COVID-19: Assessing the risks and identifying needed reforms. *JAMA, 323*(19), 1901–1902. https://doi.org/10.1001/jama.2020.5442

Colburn, B. (2022). Disability-based arguments against assisted dying laws. *Bioethics, 36*(6), 680–686. https://doi.org/10.1111/bioe.13036

Degener, T. (2024). The human rights model of disability in times of triage. *Scandinavian Journal of Disability Research, 26*(1). https://doi.org/10.16993/sjdr.1088

Deshmukh, A. A., Chen, A., Sonawane, K. B., & Cantor, S. B. (2013). The use of DALYs and QALYs in cost-effectiveness analysis. *Value in Health, 16*(3), A270. https://doi.org/10.1016/j.jval.2013.03.1389

Dunn, K., Hughes-McCormack, L., & Cooper, S. A. (2018). Hospital admissions for physical health conditions for people with intellectual disabilities: Systematic review. *Journal of Applied Research in Intellectual Disabilities, 31*, 1–10. https://doi.org/10.1111/jar.12360

Emanuel, E. J. (1994). Euthanasia: Historical, ethical, and empiric perspectives. *Archives of Internal Medicine, 154*(17), 1890–1901. https://doi.org/10.1001/archinte.1994.00420170022003

Engelman, A., Craig, L., & Iles, A. (2022). Global disability justice in climate disasters: mobilizing people with disabilities as change agents: Analysis describes disability justice in climate emergencies and disasters, mobilizing people with disabilities as change agents. *Health Affairs, 41*(10), 1496–1504.

Fineman, M. A. (2012). Elderly as vulnerable: Rethinking the nature of individual and societal responsibility. *Elder Law Journal, 20*, 71.

FitzGerald, C., & Hurst, S. (2017). Implicit bias in healthcare professionals: A systematic review. *BMC Medical Ethics, 18*(1), 19. https://doi.org/10.1186/s12910-017-0179-8

Fontalis, A., Prousali, E., & Kulkarni, K. (2018). Euthanasia and assisted dying: What is the current position and what are the key arguments informing the debate? *Journal of the Royal Society of Medicine, 111*(11), 407–413. https://doi.org/10.1177/0141076818803452

Gill, C. J. (2010). No, we don't think our doctors are out to get us: Responding to the straw man distortions of disability rights arguments against assisted suicide. *Disability and Health Journal, 3*(1), 31–38.

Göttert, E. A. F. (2025). What is fair? Ethical analysis of triage criteria and disability rights during the COVID-19 pandemic and the German legislation. *Journal of Medical Ethics, 51*(2), 139–143. https://doi.org/10.1136/jme-2023-109326

Goulden, A., Kattari, S. K., Slayter, E. M., & Norris, S. E. (2023). 'Disability is an art. It's an ingenious way to live': Integrating disability justice principles and critical feminisms in social work to promote inclusion and anti-ableism in professional praxis. *Affilia, 38*(4), 732–741. https://doi.org/10.1177/08861099231188733

Gupta, A. K., & Bansal, D. (2023). Euthanasia—Review and update through the lens of a psychiatrist. *Industrial Psychiatry Journal, 32*(1), 15–18. https://doi.org/10.4103/ipj.ipj_259_21

Hallyburton, A. (2022). Diagnostic overshadowing: An evolutionary concept analysis on the misattribution of physical symptoms to pre-existing psychological illnesses. *International Journal of Mental Health Nursing, 31*(6), 1360–1372. https://doi.org/10.1111/inm.13034

Heyer, K. (2011). Rejecting rights: The disability critique of physician assisted suicide. In *Special issue social movements/legal possibilities* (pp. 77–112). Emerald Group Publishing Limited. https://doi.org/10.1108/S1059-4337(2011)0000054007

Holness, W. (2014). Equal recognition and legal capacity for persons with disabilities: Incorporating the principle of proportionality. *South African Journal on Human Rights, 30*(2), 313–344. https://doi.org/10.1080/19962126.2014.11865111

Huang, L. L. (2025). Medical futility and the ethics of continuing treatment: A hermeneutic inquiry into patient and physician perspectives. *Philosophy, Ethics, and Humanities in Medicine, 20*(1), 33. https://doi.org/10.1186/s13010-025-00200-3

Iezzoni, L. I., Rao, S. R., Ressalam, J., Bolcic-Jankovic, D., Agaronnik, N. D., Donelan, K., et al. (2021). Physicians' perceptions of people with disability and their health care: Study reports the results of a survey of physicians' perceptions of people with disability. *Health Affairs, 40*(2), 297–306. https://doi.org/10.1377/hlthaff.2020.01452

International Bioethics Committee. (2013). *The Principle of Respect for Human Vulnerability and Personal Integrity: Report of the International Bioethics Committee of UNESCO (IBC)*. https://unesdoc.unesco.org/ark:/48223/pf0000219494

Jagoe, C., Toh, P. Y. N., & Wylie, G. (2025). Disability and the risk of vulnerability to human trafficking: An analysis of case law. *Journal of Human Trafficking, 11*(2), 220–234. https://doi.org/10.1080/23322705.2022.2111507

Javaid, A., Nakata, V., & Michael, D. (2019). Diagnostic overshadowing in learning disability: Think beyond the disability. *Progress in Neurology and Psychiatry, 23*(2), 8–10. https://doi.org/10.1002/pnp.531

Joebges, S., & Biller-Andorno, N. (2020). Ethics guidelines on COVID-19 triage—An emerging international consensus. *Critical Care, 24*(1), 201. https://doi.org/10.1186/s13054-020-02927-1

Kaundinya, T., & Schroth, S. (2022). Dismantle ableism, accept disability: Making the case for anti-ableism in medical education. *Journal of Medical Education and Curricular Development, 9*, 23821205221076660.

Kono, M., Arai, N., & Takimoto, Y. (2023). Identifying practical clinical problems in active euthanasia: A systematic literature review of the findings in countries where euthanasia is legal. *Palliative & Supportive Care, 21*(4), 705–713. https://doi.org/10.1017/S1478951522001699

Kopar, P. K., Visani, A., Squirrell, K., & Brown, D. E. (2022). Addressing futility: A practical approach. *Critical Care Explorations, 4*(7), e0706. https://doi.org/10.1097/CCE.0000000000000706

Leever, M. G., Richter, K., Nelson, P., Allman, C. J., & Wyeth, D. (2012, June). The case of do-not-resuscitate (DNR) orders and the intellectually disabled patient. In *HEC forum* (Vol. 24(2), pp. 83–90). Springer Netherlands. https://doi.org/10.1007/s10730-011-9166-5

Meltzer, L. S., & Huckabay, L. M. (2004). Critical care nurses' perceptions of futile care and its effect on burnout. *American Journal of Critical Care: An Official Publication, American Association of Critical-Care Nurses, 13*(3), 202–208.

Moreno, J. D. (1995). *Deciding together: Bioethics and moral consensus*. https://philpapers.org/rec/MORDTB

Mousavi, S. (2024). Global ethical principles in healthcare networks, including debates on euthanasia and abortion. *Cureus, 16*(4). https://doi.org/10.7759/cureus.59116

Mustaniemi-Laakso, M., Katsui, H., & Heikkilä, M. (2023). Vulnerability, disability, and agency: Exploring structures for inclusive decision-making and participation in a responsive state. *International Journal for the Semiotics of Law-Revue Internationale de Sémiotique Juridique, 36*(4), 1581–1609. https://doi.org/10.1007/s11196-022-09946-x

Ntseke, S., Coetzee, I., & Heyns, T. (2023). Moral distress among critical care nurses when executing do-not-resuscitate (DNR) orders in a public critical care unit in Gauteng. *Southern African Journal of Critical Care, 39*(2), 49–53. https://hdl.handle.net/10520/ejc-m_sajcc_v39_n2_a4

Pachys, G., Kaufman, N., Bdolah-Abram, T., Kark, J. D., & Einav, S. (2014). Predictors of long-term survival after out-of-hospital cardiac arrest: The impact of Activities of Daily Living and Cerebral Performance Category scores. *Resuscitation, 85*(8), 1052–1058. https://doi.org/10.1016/j.resuscitation.2014.03.312

Paratz, E. D., Nehme, E., Burton, S., Heriot, N., Bissland, K., Rowe, S., et al. (2023). Resuscitation of people with a do-not-resuscitate order: When does it happen and what are the outcomes? *Resuscitation, 193*, 110027. https://doi.org/10.1016/j.resuscitation.2023.110027

Parley, F. F. (2011). What does vulnerability mean? *British Journal of Learning Disabilities, 39*(4), 266–276. https://doi.org/10.1111/j.1468-3156.2010.00663.x

Peterson, A., Karlawish, J., & Largent, E. (2021). Supported Decision Making With People at the Margins of Autonomy. *The American Journal of Bioethics : AJOB, 21*(11), 4–18. https://doi.org/10.1080/15265161.2020.1863507

Phillips, K. G., England, E., & Wishengrad, J. S. (2021). Disability-competence training influences health care providers' conceptualizations of disability: An evaluation study. *Disability and Health Journal, 14*(4), 101124. https://doi.org/10.1016/j.dhjo.2021.101124

Picón-Jaimes, Y. A., Lozada-Martinez, I. D., Orozco-Chinome, J. E., Montaña-Gómez, L. M., Bolaño-Romero, M. P., Moscote-Salazar, L. R., et al. (2022). Euthanasia and assisted suicide: An in-depth review of relevant historical aspects. *Annals of Medicine and Surgery, 75*, 103380. https://doi.org/10.1016/j.amsu.2022.103380

Rand, L. Z., & Kesselheim, A. S. (2021). Controversy over using quality-adjusted life-years in cost-effectiveness analyses: A systematic literature review: Systematic literature review examines the controversy over the use of quality-adjusted life-year in cost-effectiveness analyses. *Health Affairs, 40*(9), 1402–1410. https://doi.org/10.1377/hlthaff.2021.00343

Raoofi, N., Raoofi, S., Jalali, R., et al. (2021). The worldwide investigating nurses' attitudes towards do-not-resuscitate order: A review. *Philosophy, Ethics, and Humanities in Medicine, 16*, 5. https://doi.org/10.1186/s13010-021-00103-z

Rolak, S., Elhawary, A., Diwan, T., & Watt, K. D. (2024). Futility and poor outcomes are not the same thing: A clinical perspective of refined outcomes definitions in liver transplantation. *Liver Transplantation, 30*(4), 421–430.

Saettele, A., & Kras, J. (2013). Current attitudes of anesthesiologists towards medically futile care. *Open Journal of Anesthesiology, 3*(4), 207–213.

Salles, A., Awad, M., Goldin, L., Krus, K., Lee, J. V., Schwabe, M. T., & Lai, C. K. (2019). Estimating implicit and explicit gender bias among health care professionals and surgeons. *JAMA Network Open, 2*(7), e196545–e196545. https://doi.org/10.1001/jamanetworkopen.2019.6545

Schneider, P. (2022). The QALY is ableist: On the unethical implications of health states worse than dead. *Quality of Life Research, 31*(5), 1545–1552. https://doi.org/10.1007/s11136-021-03052-4

Scopetti, M., Morena, D., Padovano, M., Manetti, F., Di Fazio, N., Delogu, G., et al. (2023, May). Assisted suicide and euthanasia in mental disorders: Ethical positions in the debate between proportionality, dignity, and the right to die. *Healthcare, 11*(10), 1470. https://doi.org/10.3390/healthcare11101470

Shildrick, M. (2008). Deciding on death: Conventions and contestations in the context of disability. *Bioethical Inquiry, 5*, 209–219. https://doi.org/10.1007/s11673-007-9074-1

Snipstad, Ø. I. M. (2022). Concerns regarding the use of the vulnerability concept in research on people with intellectual disability. *British Journal of Learning Disabilities, 50*(1), 107–114. https://doi.org/10.1111/bld.12366

Stainton, T. (2022). *Disability and assisted suicide: Elucidating some key concerns.*

Taccone, F. S. (2023). Ableism in the intensive care unit. *Intensive Care Medicine, 49*(7), 898–899. https://doi.org/10.1007/s00134-023-07084-x

Taheri, E., Wang, C., & Doost, E. Z. (2023). Emergency decision-making under an uncertain time limit. *International Journal of Disaster Risk Reduction, 95*, 103832. https://doi.org/10.1016/j.ijdrr.2023.103832

Trimble, M. (2025). First steps down the slippery slope?: An analysis of the slippery-slope argument and its application to the question of assisted suicide. *The Ulster Medical Journal, 94*(1), 42–46.

Vela, M. B., Erondu, A. I., Smith, N. A., Peek, M. E., Woodruff, J. N., & Chin, M. H. (2022). Eliminating explicit and implicit biases in health care: Evidence and research needs. *Annual Review of Public Health, 43*(1), 477–501. https://doi.org/10.1146/annurev-publhealth-052620-103528

White, D. B., & Lo, B. (2020). A framework for rationing ventilators and critical care beds during the COVID-19 pandemic. *JAMA, 323*(18), 1773–1774. https://doi.org/10.1001/jama.2020.5046

Wong, J., Duane, P. G., & Ingraham, N. E. (2020). A case series of patients who were do not resuscitate but underwent cardiopulmonary resuscitation. *Resuscitation, 146*, 145–146. https://doi.org/10.1016/j.resuscitation.2019.11.020

Chapter 5
Reproductive Autonomy, Abortion, and Disability Rights

Abstract The main driving forces behind society's changing views on autonomy, disability, and reproductive rights are technological advances, evolving legal systems, and global ethical debates. The changes in these concepts must be understood within a background of a long history of repressive policies, eugenic ideologies, and constant barriers that excessively affect people with disabilities. This chapter addresses the issues of fertility discrimination, limited access to reproductive healthcare, and ableism, which are intertwined with the policies on prenatal testing and abortion to assess their combined negative impact on reproductive justice. It brings the historical background of forced sterilization, the conflicts between the pro-choice movement and the disabled rights, and the social biases that restrict the parenting chances of the disabled population into the discussion. The chapter critically analyses the situation of the legal safeguards for the unborn with disabilities and the institutional disadvantages that disabled people have, advocating for a rights-based and inclusive approach that acknowledges the right to choose and guarantees equal reproductive decision-making for all.

Keywords Reproductive justice · Informed consent · Eugenics · Ableism · Assisted reproductive technology

5.1 Introduction

Laws, politics, philosophy, and religious precepts have all influenced the development, understanding, and practical application of various autonomy ideas. The bioethical principle states that an individual's autonomy and right to control their own life must be respected (Nepal et al., 2023). An individual's technological, social, and psychological ability to freely make decisions about matters related to their personal interests is known as autonomy. The significance of women's autonomy in healthcare decision-making for both human rights and healthcare outcomes has led to increased research on this topic (Idris et al., 2023).

A. Shibi Anilkumar, R. Veerabathiran, *Bioethics and Disability*, SpringerBriefs in Modern Perspectives on Disability Research,
https://doi.org/10.1007/978-981-95-8197-9_5

Global discrepancies in women's medical decision-making continue to exist despite appeals for women's empowerment. Females' autonomy has a substantial impact on health-related outcomes, increasing adulthood survival, treatment, and healthcare visits. Reducing morbidity and death rates in mothers and their offspring can be achieved by empowering women in healthcare decision-making and ensuring optimal use (Idris et al., 2023). For women to attain reproductive independence and rights, they must have reproductive autonomy (Loll et al., 2021).

Hailed as one of the most significant developments in reproductive politics in recent memory, reproductive justice is a cross-disciplinary feminist framework grounded in intersectionality and human rights. This radical perspective emerged from an understanding of how important it is to view sexual and reproductive concerns as highly political, profoundly social, and a significant platform for the oppression of women in particular and, later, other socially marginalized groups (Morison, 2021). The significance and obstacles to reproductive and physical autonomy, the importance of structural injustices and the social factors of health, and the necessity of paying more attention to gender and racial fairness are all topics of ongoing and growing concern (Dehlendorf et al., 2021). The core of reproductive justice is the human right to uphold one's own bodily autonomy, the decision of having children, and to raise the children we do have, in spaces that are secure and sustainable (Dehlendorf et al., 2021).

Reproductive autonomy is hampered by historical discrimination and coercive reproductive practices, inaccessible facilities, and a lack of culturally appropriate care, all of which are imposed on the disability community by nonmembers. The development of inclusive reproductive policy and healthcare provisions has also been hampered by social and economic marginalization as well as a lack of representation from the disability population (Fletcher et al., 2023). Women with disabilities were mainly excluded from narratives and discourses, although there were notable mainstream feminist movements in the twentieth century concerning pregnant women's "right to choose" abortions. Through an eugenic state agenda, regulations, and jurisprudence, disability was not only used as a justification to deny disabled people the freedom to choose an abortion, but it also became a justification for nearly unambiguous approval of abortions based on identified abnormalities in fetuses (Jain & Sengupta, 2021).

For a very long time, women have fought for women's rights to bodily integrity and self-determination. Disabled feminists argue that both movements have failed to address the issue of reproductive rights for women with disabilities as a central point of debate (Kallianes & Rubenfeld, 1997). The focus of researchers, decision-makers, and practitioners has been mainly on the fact that the limitations have a greater impact on disabled people, who are the most vulnerable group. In the past, frameworks on reproductive health and rights have not placed these groups at the center of clinical and policy decision-making (Hassan et al., 2023). Individuals with disabilities are more likely to become pregnant unintentionally, be disproportionately exposed to sexual violence, and have higher rates of maternal and newborn death and morbidity (Hassan et al., 2023).

Forcible sterilizations and reproductive experiments became common in the nineteenth century. In contrast, the twentieth century was marked by the continued harsh treatment of "the feeble-minded," a term that was often applied to people with developmental disabilities, as a result of several policies (Brennen, 2023). This chapter investigates the interconnection between reproductive freedom, abortion, and disability rights, emphasizing the necessity to mediate autonomy, rights, and justice for disabled people. Furthermore, the chapter reveals how historical discrimination, restrictive policies, and social biases affect reproductive choices and access to care, and it also stresses the necessity of a comprehensive, rights-based framework capable of safeguarding bodily integrity and securing equal reproductive decision-making for disabled persons.

5.2 Discrimination in Fertility Services

Infertility is one of the most prevalent chronic illnesses among individuals of child-bearing age, affecting reproductive-age couples globally, with about equal numbers of instances occurring in men and women. Childlessness and infertility are problems of gender inequality, with the female partner bearing the majority of the burden. Women are stigmatized, shunned, and subjected to intimate partner violence or psychological abuse in many of these societies (Ombelet & Lopes, 2024). Fertility rates have decreased as a result of growing urbanization and industrialization, which have also increased the cost of childbirth and decreased labor demand (Zhang et al., 2023).

Humans have the right to reproduce, yet there are numerous possible obstacles to receiving treatment when infertility exists, which can result in inequitable access. There are many different medical approaches to treating infertility, but the most fall under the general category of "assisted reproductive technology" (ART), which is a choice for those who struggle to conceive to increase their chances of getting pregnant. In vitro fertilization (IVF) is the most popular ART method (Mackay et al., 2023).

Over time, reproductive technology users have established communities of resistance where the collective imaginary is broadened outside traditional conceptions of family foundation. An examination of ART users' view toward PwD provides an opportunity to investigate the potential outline of a "disability imaginary" in the ART framework, where the scope of its demands is made effective where PwD's are involved in policies and laws that form opportunities and futures, even though the indication of such development or struggle in the area of disability is less evident (Karpin & Mykitiuk, 2021).

Significant differences in reproductive health outcomes have probably been caused by experiences of stigma, discrimination, and exclusion combined with a medical system that does not sufficiently meet the requirements of individuals with disabilities (Biggs et al., 2023). Many disabled persons worry that if they have sex or seek reproductive healthcare from medical facilities, they will be scrutinized or be subjected to legal scrutiny. Social stigma and mental health issues (such as anxiety, despair, helplessness, etc.) are caused mainly by societal stigma, endangering their dignity and general well-being. People with disabilities face many social obstacles that impede their ability to access reproductive healthcare (Shrivastava et al., 2025).

Due to a problematic web of inequity, including unfavorable public attitudes and cultural presumptions as well as natural impediments, including laws, regulations, institutions, and services that lead to social exclusion and marginalization, people with disabilities are unable to obtain health services. People with disabilities' (PwDs') approach to sexual and reproductive health (SRH) services is further impeded by inaccessible medical facilities, insensitive healthcare professionals, a lack of awareness about disabilities, and a lack of information specific to their health requirements (Ahumuza et al., 2014).

Reproductive health issues affect women who are physically disabled. Pregnancy brings about physical and psychological changes that are extremely difficult to adapt to in the case of women with disabilities (Gül & Koruk, 2019). Many hurdles make it difficult for women with physical disabilities to get reproductive healthcare services. Some of the barriers that prevent them from accessing reproductive health services include high healthcare costs, traditional barriers, lack of family support, poor physical accessibility, lack of knowledge about reproductive healthcare services and rights, fear of inadequate care, negative attitudes of healthcare providers, concerns about confidentiality, and difficulty communicating with the providers (Haque et al., 2023).

Social exclusion has been the biggest obstacle for those with cognitive impairments. People with mental disorders were frequently institutionalized in the early decades of this century. People with cognitive disabilities still struggle to become independent outside of segregated settings, not because of personal limits but rather because they are unwanted (Block, 2000; Shanmugam et al., 2014).

For example, the infertility issue in individuals with intellectual and developmental disabilities (IDD) is more of a social and behavioral nature that hinders them from having children and having the desire to have children than a biological issue. Concerns about the reproductive rights of persons with IDD are regularly voiced by their parents, disability support workers, and medical professionals, which might either overtly or covertly limit their reproductive options. It is more the case that women suffering from IDD experience unwanted pregnancies, leading to more frequent requests for sexual education and reproductive health services to be made readily accessible. People with IDD, despite being entitled to the same sexual and reproductive rights as everybody else, still receive delayed prenatal care and less sexual education than others. Their care must be informed by a knowledge of these injustices, given the history of sterilization and stigma (Rubenstein et al., 2022).

The idea of informed choice and the objective of giving clients a wide range of contraceptive options serve as the foundation for family planning initiatives. However, a variety of obstacles, including supply and demand considerations, restrict an individual's access and actual choice. Everyone has the fundamental right to equal choice, which is essential to satisfying the wide range of customer needs.

Contraceptive choice and availability are still significantly hampered by provider bias, which includes prejudice based on client age, parity, and marital status (Solo & Festin, 2019).

Because of presumptions that women with disability are not sexually active and don't want children, research on SRH, the right to maternal health services, and further obstetrical services for women with disabilities is scarce. Individuals with disabilities are more likely than their peers without impairments to have limited access to medical facilities (Trani et al., 2011). Ableism harms people's reproductive health as well as their access and involvement with reproductive health (RH) services. Significant differences in RH results have probably been caused by discrimination, exclusion, and stigma combined with a healthcare system that does not sufficiently meet the requirements of those with disabilities (Biggs et al., 2023).

5.3 Sterilization of Disabled People—Past and Present

Although the concept is far older, Francis Galton established eugenics as a formal subject in the late 1800s. The Greek word eugenes, which means "well" and "born," is the origin of the word "eugenics" (Garver). To control selection without upsetting the populace, Plato even suggested a covert lottery. The rise of monotheistic religions hindered the spread of such views, but support for eugenics was revived and expanded in the modern era by philosophers such as Darwin and the growing authority of science (Schexnayder, 2021).

By 1890 or so, Galton had come around to the idea of "positive eugenics," which holds that society should implement policies that promote the most talented and capable people to have larger families. For many years, negative eugenics policies intended to stop specific groups from "tainting" the population were the primary focus, although several US intellectuals supported positive eugenics. Restrictive immigration regulations, which commenced with the Chinese Exclusion Act of 1877, culminated in the Immigration Restriction Act of 1924, were its most potent manifestation. This act, which was highly influenced by negative eugenics, set national quotas that severely restricted immigration from southern and eastern Europe while favoring those from northern and western Europe. For more than 40 years, it has had a significant influence on US immigration policy (Reilly, 2015).

Although less explicitly eugenic, involuntary sterilization became a common national strategy to curb population growth. In 1907, Indiana became the first state in the world to pass an imposed sterilization law, targeting specific groups and raising concerns about due process and equal protection. In 1913, the legislature rejected a proposal to sterilize persons in institutions to stop the birth of children with mental disabilities. As the state prepared to deinstitutionalize long-term inhabitants, it passed a more judiciously drafted permissive sterilization law after 12 years, which included proper process protections. Until after World War II, eugenic sterilization programs in the United States occurred about entirely within state institutions housing those labeled "mentally defective" or "insane." By the late 1930s, these programs began to decline. In Germany, the history of eugenic sterilization began in 1934, when the Nazi government implemented a sweeping "racial hygiene" law (Reilly, 2015).

The allegations of forced female sterilizations have their roots in the era of World War II, when the Nazis performed sterilizations on Jewish, Roma, and Sinti people. Also, the same treatment was carried out by the Imperial Japanese Army in Korea and the Indian Health Service in conjunction with certain doctors who targeted Native American women for sterilization. The end of World War II in 1945 did not put an end to the worldwide reports of coerced sterilization. In Canada, there are still ongoing investigations, while South Africa is looking into cases of women living with HIV in the KwaZulu-Natal and Gauteng areas. These practices are still going on, and, therefore, the informed-consent process for female sterilization needs to be reviewed very carefully, and its ethical implications should be addressed (Maila et al., 2025).

Female sterilization, in which tubal ligation is a lasting technique of birth control that entails obstructing the fallopian tubes to prevent pregnancy. Just like any other surgical procedure, female sterilization requires consent based on the information being provided. Since female sterilization is irretrievable, women must go through extensive counseling about the procedure and its risks, benefits, and alternatives, and this counseling should be duly documented. Such human rights provisions underscore the necessity of obtaining informed consent before carrying out such operations to ensure they are truly voluntary and not coerced (Maila et al., 2025).

Institutions used more than simply intimidation and threats to coerce sterilizations. The harsh, antiquated sterilization rules either eliminated the ability to give consent or permitted forced methods. Specific laws required individuals with ID to appear before a committee with the authority to make decisions; some of these had an appeals procedure, but the appeals' success rate was often quite low. Vulnerable people were frequently the targets of cruel operations that disregarded their right to bodily autonomy. The consent procedure occasionally included coercive incentives (Rowlands & Amy, 2019).

The right to have children is a fundamental human right supported by legal frameworks worldwide, including privacy rights, and explicitly recognized in international treaties and national constitutions. The 1994 UN Conference on Population and Development in Cairo stated that governments must protect individuals' reproductive rights as their primary duty. Forced sterilization, because of its severe and very adverse effects on mental and social health, has been acknowledged by the UN Human Rights Committee as a violation of the right not to be free from torture, cruel, inhuman, or degrading treatment. It is so discriminating and harmful that it falls within the international definition of torture. The UN Special Rapporteur on the Right to Health adds that laws that allow medical interventions like sterilization to occur without informed consent infringe upon bodily and psychological integrity and may even amount to torture or ill-treatment (Iemelianenko et al., 2019).

Higher levels of self-determination among adults with disabilities were associated with more financial independence, independent living, higher-paying jobs, and frequent promotions. But by definition, guardianship changes a person with a disability's ability to exercise self-determination and make independent decisions. Many of the rights that the rest of us enjoy are also denied by this formal system of substituted decision-making, which can deprive them of the chance to make a variety of decisions about their lives. This procedure, regulated by state law, involves a court designating a guardian to make decisions on behalf of a ward or respondent who is either completely or partially incompetent or otherwise lacking capacity (Jameson et al., 2015).

All people with disabilities are recognized as having legal capacity in all aspects of life under Article 12 of the CRPD. Some people with intellectual disability (ID) would rather have decisions about their lives made on their behalf. Parents of people with ID believed that it was essential that they act as their children's guardians. So there are suggestions for enhancing the legal capacity of people with impairments (Werner & Chabany, 2016).

Since the state lays down the rules and maintains the social order, people usually think there is no alternative but to accept; otherwise, they will face penalties. This results in a form of coercion, as the public begins to believe that the only way out is to comply. In the medico-legal context, disabled individuals may have no real choice regarding institutionalization or the use of pharmaceutical treatments. Through the stigma attached to disability labels, government and medical authorities may pressure individuals simply because the law permits such actions, even when these actions may have adverse effects on the person. Limiting resources and restricting access further enable this dynamic, making control and coercion easier to carry out (De Vera, 2023).

In the field of mental health, the use of coercive measures like mechanical restraints raises ethical questions that need theoretical justification in cases of high vulnerability. The employment of coercive measures is guided by core moral values, such as dignity, autonomy, beneficence and nonmaleficence, fairness, and vulnerability, which also promote a deeper comprehension of the underlying principles of legal frameworks. These values are regarded as prima facie, meaning they typically have equal moral weight but may be given greater weight in particular circumstances (Ramos-Pozón et al., 2025).

5.4 Pro-Choice Versus Disability Rights Tensions

Feminism and the occasionally opposing pro-choice movement are the foundations of the contemporary reproductive rights movement. The pro-choice stance is mostly in line with feminist beliefs, although not every feminist supports the idea of abortion. The pro-choice philosophy states that a pregnant woman has the right to control their body and make decisions about giving birth or not. This is stated in the popular slogan "my body, my rules," which represents the pro-choice attitude. Those in politics who support the pro-choice viewpoint argue that providing access to abortion and other reproductive technologies is even more critical than the actual procedure. The pro-choice feminist reproductive movement, which echoes the rhetoric of individuality and freedom of choice, permits individuals to choose the kind of children they have (Moor, 2019).

It is unethical and has historically eugenic overtones to support legal abortion only in cases of fetal impairment. Pro-choice people must take into account the moral issues that arise even in campaigns to amend anti-abortion legislation (Masters, 2017). Raising the threat of antenatal diagnosis of possible or confirmed inherited disabilities as justification for providing way to abortion in the third trimester has been the main tactic used by national pro-choice organizations. Disregarding family support program agendas and their crucial role in miscarriage debates, erasing the challenges of accessing abortion services, and generalizing disability as disastrous, pitiful, or inspirational are all incompatible with the typical increasing values of the pro-choice and disability rights communities.

In their defense of the right to abort these pregnancies, pro-choice advocates frequently promote financing for social services and support initiatives that cater to the needs of all households, including those with children with disabilities. Some pro-choice organizations have told tales of families having to make the "excruciating choice" to end a desired pregnancy because their fetus has been diagnosed with a disability in an effort to humanize difficult late abortion decisions. The pro-choice movement's best chance to change the abortion debate from a medical argument about fetal suffering and viability to a positive, potentially revolutionary platform on human rights in general is provided by the disability rights framework. It helps to have a functional definition of impairment to understand this (Jesudason & Epstein, 2011).

While the broader ethical issues relating to antenatal testing for disability are not novel, and have been the focus of wide analysis, other elements for abortion in the regulation are built on a health risk, whereas the disability ground is the only one that is based on the child's, or future child's, features. The majority of abortions are also subject to a time limit such that they can only take place in the first 24 weeks of gestation. But, there are, in fact, many diverse options for eliminating the negative message in the law caused by the disability ground, and not all of these would require a restriction on access to abortion. More liberal changes to the law, for example, could eradicate the exact stigmatizing provision without restricting abortion access (Robinson, 2023a, 2023b).

The availability of abortion for the reason that the child will be disabled must be understood in the broader context of the widespread practice of prenatal testing for disability. Currently, prenatal screening is offered by the National Health Service and is available for a wide range of conditions. Although the testing is optional, it is a routine practice during pregnancy. One major critique of prenatal testing and abortion on the grounds of disability is the expressivist argument, according to which these practices express discriminatory or hurtful views about people living with the same conditions that can be detected through prenatal testing, and send the message that their lives are less valuable, or that they would be better off dead (Robinson, 2023a, 2023b).

One of the biggest obstacles to the equality of individuals with disabilities is negative views. Governments and organizations have responded by introducing initiatives to change people's perceptions. Young people and educated people in general have a more positive attitude toward disabled people. Negative views based on misunderstanding, ignorance, or lack of knowledge not only manifest themselves in direct or indirect discrimination but also in open demonization. Nongovernmental organizations, advocacy groups, and governments can formulate policies at both the individual and organizational levels. The CRPD specifies these rights, and it is the duty of states to pass laws to maintain and protect them, which includes changing public attitudes (Fisher & Purcal, 2017).

The inclusion of people with disabilities in society is still affected by the prejudices that are deeply rooted in culture. Stigma not only hinders the process of identifying, diagnosing, treating, and managing disabilities but also contributes to a cycle of exclusion and injustice by perpetuating stereotypes and misconceptions. There have been no efforts to raise awareness and educate the public to promote social inclusion in the community, even though strong policies are in place. The provision of necessary technologies, equal access to jobs and education, and the creation of a barrier-free environment are all essential for the realization of inclusivity and equity. Nevertheless, limited access and lack of funds for this technology are often cited as major hindrances to its use. These will necessitate a significant change in perspective and focus on policy. This cannot be accomplished without involving individuals with disabilities in the formulation of policies and decision-making processes (The Lancet Regional Health-Southeast Asia, 2025).

5.5 Legal Protections for Fetuses with Disabilities

The 1950s saw the development of accurate human karyotyping, which made the identification of chromosomal aberrations easy and, hence, led to widespread screening, particularly for Down syndrome, which became systematic in the 1990s. Currently, various methods can be used to screen and diagnose a wide range of medical conditions, among which is cell-free fetal DNA testing, which provides results as early as 9 weeks of pregnancy (Giric, 2016).

By 2018, at least twelve states had passed some selective abortion prohibition, which forbade a doctor from performing an abortion on a woman if her motivation was based on the fetus's sex, race, or genetic impairment. While some states have few challenges to the validity of these selective abortion bans, other courts have declared the prohibitions unlawful or temporarily halted their execution, and more legal action is still ongoing (Gooder, 2018).

While some bans are unique to particular disability, others were implemented as part of larger prohibitions of abortions seen to be discriminatory. Bans with a broader reach could be presented as more comprehensive anti-discrimination initiatives. Current disability-based prohibitions also emphasize how the prohibitions affect children with disabilities rather than pregnant individuals with disabilities. For the most part, these criticisms align with the argument made here, which is that disability-based prohibitions ignore crucial elements of reproductive justice for those who wish to get pregnant, are already pregnant, or wish to have children. However, focusing only on the import of disability-based prohibitions for children with disabilities, or even for individuals with disabilities in general, runs the risk of marginalizing those most impacted by abortion prohibitions: parents or those who may be pregnant (Francis, 2023).

Abortion is legal on one or more grounds in nearly all countries, usually as exceptions to criminal law. Canada is the only country that has fully decriminalized abortion, following a 1988 Supreme Court decision. No other nation has removed abortion entirely from its legal framework. In the United Kingdom, abortion is still considered a crime, according to the 1861 Offences Against the Person Act and the 1929 Infant Life Preservation Act, despite the 1967 Abortion Act and its 1990 amendments, which allowed the procedure under particular conditions (Berer, 2017).

India legalized abortion in 1971, but the lack of proper implementation has resulted in continued morbidity and mortality; even 15 years ago, clinic registration was still complicated. Cuba has provided access to abortion on demand through its national health system for up to 10 weeks of pregnancy since 1965, which makes it the only country in this regard throughout Latin America and the Caribbean. Japan's 1948 abortion law, initially rooted in eugenics, functioned as a liberal policy in practice and made abortion the primary form of birth control; references to eugenics were removed in a 1996 reform (Berer, 2017).

Restrictive regulations, political speeches, and the general public's opinions show anti-abortion sentiment. In a few of these Western countries, like Poland and the United States, abortion rights were lessened to a large degree. On the other hand, Iceland was able to support reproductive freedom by passing more liberal laws that raised the legal abortion limit from 16 to 22 weeks in 2019. As there was an objective behind it, conservative politicians did not just oppose that measure but rather kept reiterating the belief that abortion is bad no matter what (Johnson et al., 2025).

Traditional feminist ideals that view parenting as essential to a woman's identity are directly associated with abortion stigma and regret-related anxieties. These standards put pressure on women to fit into socially mandated roles, which frequently limit their reproductive options and undermine their autonomy (Johnson et al., 2025).

5.6 Parenting with Disabilities: Societal Assumptions

It is rare for disabled men and women to be viewed as parents. Conversely, they have been deterred and kept from becoming parents. Parents with impairments, both men and women, raise their children on their own, sometimes with help from personal assistants. However, mothers with impairments struggle to be seen as mothers or women since they are largely enmeshed in the rhetoric of disability. Women with disabilities have typically been viewed as caregivers rather than as partners or parents who look after others (Selander & Engwall, 2021).

Adoption of children by disabled parents is widespread throughout the world. Disabled parents are at a higher risk of losing custody and having their parental rights taken away. They usually experience many hardships in the process of adoption, including prejudice, access barriers, systemic misconceptions, inadequate child welfare, and social workers and assessors who often lack appropriate training and understanding (Rosenberg-Lavi & Herbst-Debby, 2025). Figure 5.1 highlights major structural, social, and institutional barriers that limit reproductive autonomy for persons with disabilities and social outcomes.

New research reveals that the disability community still has to deal with significant obstacles in having and raising children, one of those being the continued legacy of forced or coerced sterilization. Although studies have pointed out different ways to parent, the focus has been mainly on techniques related to reproductive technologies, such as in vitro fertilization (IVF). The Victorian Law Reform Commission's 2007 report on parenting laws, for example, listed disability as a significant barrier to IVF but gave only a brief account of adoption and suggested same-sex attraction, not disability, as the main hurdle. Although Article 23 of the CRPD explicitly protects the right of persons with disabilities to adopt, there has been little examination of whether signatory states have translated these obligations into effective domestic law (Connell, 2017).

Persistent monitoring of individuals with disabilities has been interpreted as fulfilling a social or public obligation to provide care for these individuals (Ho et al., 2014). Continuous monitoring of people with disabilities has often been framed as

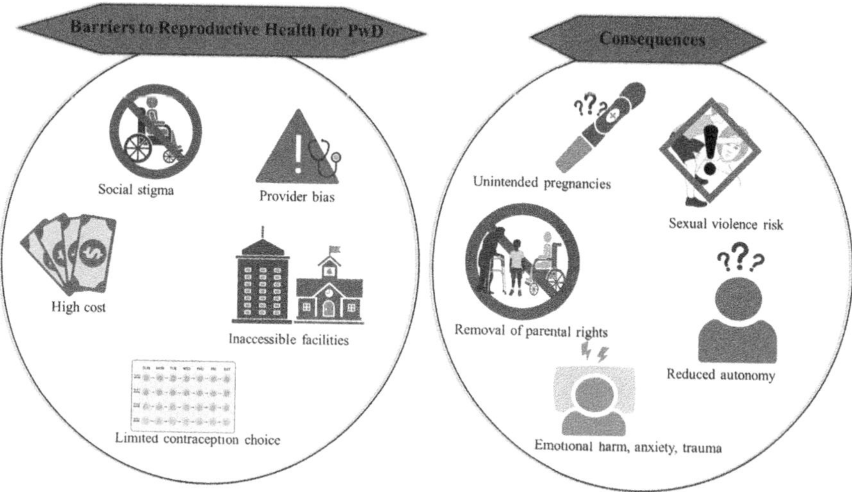

Fig. 5.1 Barriers to reproductive health for persons with disabilities and their associated consequences. The figure illustrates barriers to reproductive health for persons with disabilities (PwD) on the left and the consequences of these barriers on the right. The barrier section includes icons representing inaccessibility, medical bias, financial barriers, lack of institutional support, and limited access to reproductive health information or contraception. The consequences section shows icons representing unintended pregnancy, vulnerability to abuse, denial of reproductive autonomy, psychological distress, and confusion or lack of guidance. The two sections are visually divided: the barrier icons are placed under a red banner labeled "Barriers to Reproductive Health for PwD," and the consequences are grouped in a large circle under a banner labeled "Consequences"

a collective or public responsibility to "care" for them. Consequently, parents who have disabilities face an unequal share of involvement with the child welfare system when compared with nondisabled parents. It is a commonly accepted fact that children born to parents suffering from intellectual or mental health disabilities are more likely to be removed from their homes compared to the rest of the kids. Such trends are further supported by the widespread belief that disabled parents are always unfit to provide proper care. These beliefs can push child welfare agencies and courts to make early decisions that these parents will not gain from the supportive services (Albert & Powell, 2021).

To really assist families with parents who are disabled, child welfare professionals need to be aware of the strategies that parents adapt, suitable ways of evaluating the capacity of parents, the support they can get from disability, the interventions that are based on evidence, and the laws regarding disability rights that are applicable. Moreover, parents with disabilities frequently interact with multiple service providers, leading to greater exposure and greater risk of "surveillance bias." Existing research has also been limited, focusing mainly on disability-specific groups such as parents with intellectual or psychiatric disabilities, although emerging research suggests parents involved with the child welfare system often have coexisting disabilities (Albert & Powell, 2021).

Social policy has historically controlled the parenting practices of individuals with disabilities in several ways, such as institutionalization, forced sterilization, and the termination of parental rights. People with disabilities are now more included in community, work, and educational settings, and the general policy landscape has evolved in many ways. Parents with disabilities may be subject to increased scrutiny from child protection agencies because many child welfare procedures imply that a parental handicap is a high risk for abuse. States should review their child welfare laws for language related to disability in their evaluations and other procedures, not just when ending parental rights (Lightfoot et al., 2010).

5.7 Conclusion

The concepts of reproductive autonomy and disability rights should be viewed as principles that support each other, and, at the same time, an ethical, legal, and clinical framework grounded in equality and justice is needed. Historical eugenics, institutional coercion, discriminatory healthcare systems, and entrenched stigma have been observed as evidence that these barriers still exist in present times and have influenced reproductive environments. The existence of such obstacles has restricted access to fertility services, complicated abortion debates, and disabled parents have been subjected to disproportionately high scrutiny. Justice needs to be reinforced through informed consent, removal of ableist assumptions in reproductive policies, and more accessibility in healthcare. Also, it is necessary to allow disabled individuals to participate meaningfully in all decision-making areas. Disability justice in pro-choice discussions offers a more comprehensive view of reproductive rights, a vision that not only protects but also expands possibilities and recognizes the worth of everyone. Having such a framework is very important for creating fair systems in which all human beings can equally and freely exercise their reproductive rights.

References

Ahumuza, S. E., Matovu, J. K., Ddamulira, J. B., & Muhanguzi, F. K. (2014). Challenges in accessing sexual and reproductive health services by people with physical disabilities in Kampala, Uganda. *Reproductive Health, 11*(1), 59. https://doi.org/10.1186/1742-4755-11-59

Albert, S. M., & Powell, R. M. (2021). Supporting disabled parents and their families: Perspectives and recommendations from parents, attorneys, and child welfare professionals. *Journal of Public Child Welfare, 15*(5), 529–529. https://doi.org/10.1080/15548732.2020.1751771

Berer, M. (2017). Abortion law and policy around the world: In search of decriminalization. *Health and Human Rights, 19*(1), 13–27.

Biggs, M. A., Schroeder, R., Casebolt, M. T., Laureano, B. I., Wilson-Beattie, R. L., Ralph, L. J., et al. (2023). Access to reproductive health services among people with disabilities. *JAMA Network Open, 6*(11), e2344877. https://doi.org/10.1001/jamanetworkopen.2023.44877

Block, P. (2000). Sexuality, fertility, and danger: Twentieth-century images of women with cognitive disabilities. *Sexuality and Disability, 18*(4), 239–254. https://doi.org/10.1023/A:1005642226413

Brennen, B. E. (2023). *Advocating for policy guaranteeing the right to receive sterilization.* Doctoral dissertation. University of Pittsburgh.

Connell, B. (2017). Some parents are more equal than others: Discrimination against people with disabilities under adoption law. *Laws, 6*(3), 15. https://doi.org/10.3390/laws6030015

De Vera, P. (2023). The control and coercion of disabled individuals: The injustice of legal violence of medical and pharmaceutical interventions. *Crossing Borders: Student Reflections on Global Social Issues, 5*(2). https://doi.org/10.31542/kqrk6h08

Dehlendorf, C., Akers, A. Y., Borrero, S., Callegari, L. S., Cadena, D., Gomez, A. M., Hart, J., Jimenez, L., Kuppermann, M., Levy, B., Lu, M. C., Malin, K., Simpson, M., Verbiest, S., Yeung, M., & Crear-Perry, J. (2021). Evolving the Preconception Health Framework: A Call for Reproductive and Sexual Health Equity. *Obstetrics and Gynecology, 137*(2), 234–239. https://doi.org/10.1097/AOG.0000000000004255

Fisher, K. R., & Purcal, C. (2017). Policies to change attitudes to people with disabilities. *Scandinavian Journal of Disability Research, 19*(2), 161–174. https://doi.org/10.1080/15017419.2016.1222303

Fletcher, J., Yee, H., Ong, B., & Roden, R. C. (2023). Centering disability visibility in reproductive health care: Dismantling barriers to achieve reproductive equity. *Women's Health, 19*, 17455057231197166. https://doi.org/10.1177/17455057231197166

Francis, L. (2023). The reproductive injustices of abortion bans for disability. *Journal of Law, Medicine & Ethics, 51*(3), 490–496. https://doi.org/10.1017/jme.2023.127

Giric, S. (2016). Strange bedfellows: Anti-abortion and disability rights advocacy. *Journal of Law and the Biosciences, 3*(3), 736–742. https://doi.org/10.1093/jlb/lsw056

Gooder, T. (2018). Selective abortion bans: The birth of a new state compelling interest. *University of Cincinnati Law Review, 87*, 545.

Gül, S., & Koruk, F. (2019). Fertility problem characteristics experienced by women with physical disability and distressing factors. *Sexuality and Disability, 37*(1), 63–75. https://doi.org/10.1007/s11195-019-09559-6

Haque, N., Sharaque, A. R., Sikder, A. S., Sultana, M., Noor, I. N., Alam, U. K., & Dastider, S. (2023). Barriers faced by women with physical disabilities for reproductive health care services utilization. *Journal of Asian Multicultural Research for Medical and Health Science Study, 4*(2), 36–45. https://doi.org/10.47616/jamrmhss.v4i2.394

Hassan, A., Hirz, A. E., Yates, L., & Hing, A. K. (2023). Rebuilding a reproductive future informed by disability and reproductive justice. *Women's Health Issues, 33*(4), 345–348. https://doi.org/10.1016/j.whi.2023.04.006

Ho, A., Silvers, A., & Stainton, T. (2014). Continuous surveillance of persons with disabilities: Conflicts and compatibilities of personal and public goods. *Journal of Social Philosophy, 45*(3).

Idris, I. B., Hamis, A. A., Bukhori, A. B. M., Hoong, D. C. C., Yusop, H., Shaharuddin, M. A. A., et al. (2023). Women's autonomy in healthcare decision making: A systematic review. *BMC Women's Health, 23*(1), 643. https://doi.org/10.1186/s12905-023-02792-4

Iemelianenko, V. V., Gornostay, A. V., & Ivantsova, A. V. (2019). Reproductive rights violations: Forced sterilization and restriction of voluntary sterilization. *Wiadomosci Lekarskie, 12*, 2536–2540. https://doi.org/10.36740/WLek201912223

Jain, D., & Sengupta, S. (2021). Reproductive rights and disability rights through an intersectional analysis. *Jindal Global Law Review, 12*(2), 337–357. https://doi.org/10.1007/s41020-021-00153-6

Jameson, J. M., Riesen, T., Polychronis, S., Trader, B., Mizner, S., Martinis, J., & Hoyle, D. (2015). Guardianship and the potential of supported decision making with individuals with disabilities. *Research and Practice for Persons with Severe Disabilities, 40*(1), 36–51. https://doi.org/10.1177/1540796915586189

Jesudason, S., & Epstein, J. (2011). The paradox of disability in abortion debates: Bringing the pro-choice and disability rights communities together. *Contraception, 84*(6), 541–543. https://doi.org/10.1016/j.contraception.2011.08.022

Johnson, M. A., Bloomer, F., & Pétursdóttir, G. M. (2025). "For some reason, she just wasn't able to have an abortion": Social attitudes, reproductive autonomy, and the taboo of regret. *Sex Roles, 91*(4), 32. https://doi.org/10.1007/s11199-025-01583-5

Kallianes, V., & Rubenfeld, P. (1997). Disabled women and reproductive rights. *Disability & Society, 12*(2), 203–222. https://doi.org/10.1080/09687599727335

Karpin, I., & Mykitiuk, R. (2021). Reimagining disability: The screening of donor gametes and embryos in IVF. *Journal of Law and the Biosciences, 8*(2), lsaa067. https://doi.org/10.1093/jlb/lsaa067

Lightfoot, E., Hill, K., & LaLiberte, T. (2010). The inclusion of disability as a condition for termination of parental rights. *Child Abuse & Neglect, 34*(12), 927–934. https://doi.org/10.1016/j.chiabu.2010.07.001

Loll, D., Fleming, P. J., Stephenson, R., King, E. J., Morhe, E., Manu, A., & Hall, K. S. (2021). Factors associated with reproductive autonomy in Ghana. *Culture, Health & Sexuality, 23*(3), 349–366.

Mackay, A., Taylor, S., & Glass, B. (2023). Inequity of access: Scoping the barriers to assisted reproductive technologies. *Pharmacy, 11*(1), 17. https://doi.org/10.3390/pharmacy11010017

Maila, S. M., Castelyn, C., & Adam, S. (2025). Informed consent and ethical issues pertaining to female sterilization—Scoping review. *International Journal of Gynecology & Obstetrics, 169*(3), 1037–1064. https://doi.org/10.1002/ijgo.16100

Masters, A. (2017). The ethics of Brazilian abortion law in the age of Zika: A pro-choice, pro-disabled-rights perspective. *Voices in Bioethics, 3*. https://doi.org/10.7916/vib.v3i.6032

Moor, A. (2019). Disrupting choice: Confronting and reexamining choice in selective abortion. *Knots: An Undergraduate Journal of Disability Studies, 4*(1).

Morison, T. (2021). Reproductive justice: A radical framework for researching sexual and reproductive issues in psychology. *Social and Personality Psychology Compass, 15*(6), e12605. https://doi.org/10.1111/spc3.12605

Nepal, A., Dangol, S. K., Karki, S., & Shrestha, N. (2023). Factors that determine women's autonomy to make decisions about sexual and reproductive health and rights in Nepal: A cross-sectional study. *PLOS Global Public Health, 3*(1), e0000832.

Ombelet, W., & Lopes, F. (2024). Fertility care in low and middle income countries: Fertility care in low-and middle-income countries. *Reproduction and Fertility, 5*(3). https://doi.org/10.1530/RAF-24-0042

Ramos-Pozón, S., Román-Maestre, B., & Blánquez, B. (2025). Coercive measures in disability and mental health care services: Mechanical restraints from a bioethical and legal perspective in Spain. *International Journal of Law and Psychiatry, 99*, 102067. https://doi.org/10.1016/j.ijlp.2024.102067

Reilly, P. R. (2015). Eugenics and involuntary sterilization: 1907-2015. *Annual Review of Genomics and Human Genetics, 16*, 351–368. https://doi.org/10.1146/annurev-genom-090314-024930

Robinson, H. (2023a). Prenatal testing, disability equality, and the limits of the law. *The New Bioethics, 29*(3), 202–215. https://doi.org/10.1080/20502877.2022.2145672

Robinson, H. (2023b). Prenatal testing, disability, and the ethical society. *The New Bioethics, 29*(3), 195–201. https://doi.org/10.1080/20502877.2023.2240173

Rosenberg-Lavi, S., & Herbst-Debby, A. (2025, May). The compulsory adoption of children of parents with disabilities in Israel: An analysis. In *Women's studies international forum* (Vol. 110, p. 103077). Pergamon. https://doi.org/10.1016/j.wsif.2025.103077

Rowlands, S., & Amy, J. J. (2019). Sterilization of those with intellectual disability: Evolution from non-consensual interventions to strict safeguards. *Journal of Intellectual Disabilities: JOID, 23*(2), 233–249. https://doi.org/10.1177/1744629517747162

Rubenstein, E., Ehrenthal, D. B., Nobles, J., Mallinson, D. C., Bishop, L., Jenkins, M. C., et al. (2022). Fertility rates in women with intellectual and developmental disabilities in Wisconsin Medicaid. *Disability and Health Journal, 15*(3), 101321. https://doi.org/10.1016/j.dhjo.2022.101321

Schexnayder, S. (2021). *Eugenics in the United States: The forgotten movement* (Vol. 5). Cheniere.

Selander, V., & Engwall, K. (2021). Parenting with assistance: The views of disabled parents and personal assistants. *Scandinavian Journal of Disability Research, 23*(1), 136–146. https://doi.org/10.16993/sjdr.775

Shanmugam, M., Shivakumar, V., Anitha, V., Meenapriya, B. P., Aishwarya, S., & Anitha, R. (2014). Behavioral pattern during dental pain in intellectually disabled children: A comparative study. *International Scholarly Research Notices, 2014*(1), 824125. https://doi.org/10.1155/2014/824125

Shrivastava, S. R., Bobhate, P. S., & Nair, N. (2025). Facilitating access of disabled persons to avail reproductive health by overcoming social barriers. *Asian Pacific Journal of Reproduction, 14*(1), 47–48. https://doi.org/10.4103/apjr.apjr_204_24

Solo, J., & Festin, M. (2019). Provider bias in family planning services: A review of its meaning and manifestations. *Global Health: Science and Practice, 7*(3), 371–385. https://doi.org/10.9745/GHSP-D-19-00130

The Lancet Regional Health-Southeast Asia. (2025). Inclusivity for people living with disability: The gap between understanding and implementing. *The Lancet Regional Health. Southeast Asia, 37*, 100614. https://doi.org/10.1016/j.lansea.2025.100614

Trani, J. F., Browne, J., Kett, M., Bah, O., Morlai, T., Bailey, N., & Groce, N. (2011). Access to health care, reproductive health and disability: A large scale survey in Sierra Leone. *Social Science & Medicine, 73*(10), 1477–1489. https://doi.org/10.1016/j.socscimed.2011.08.040

Werner, S., & Chabany, R. (2016). Guardianship law versus supported decision-making policies: Perceptions of persons with intellectual or psychiatric disabilities and parents. *American Journal of Orthopsychiatry, 86*(5), 486.

Zhang, T. T., Cai, X. Y., Shi, X. H., Zhu, W., & Shan, S. N. (2023). The effect of family fertility support policies on fertility, their contribution, and policy pathways to fertility improvement in OECD countries. *International Journal of Environmental Research and Public Health, 20*(6), 4790. https://doi.org/10.3390/ijerph20064790

Chapter 6
Assisted Reproduction, Surrogacy, and Genetic Selection

Abstract This chapter studies the multifaceted and intricate nature of the ethical, social, and disability issues that are related to assisted reproduction, mainly concentrating on genetic selection, surrogacy, and the reproductive rights of disabled people. The continuous advancement of technologies, such as IVF, PGD, PGT, and CRISPR, has brought about unprecedented control over reproduction but at the same time raised concerns about eugenics, inequality, embryo selection, and the moral status of disability. The chapter contextualizes these disputes in the frame of reproductive justice, emphasizing the role of overlapping oppressions in determining the availability of reproductive care. It deals with the present-day debates about "designer babies," the setting of severity thresholds for embryo selection, and the expressivist criticism. Additionally, it thoroughly evaluates the hurdles disabled persons encounter as intended parents or surrogates, such as medical bias, infrastructural inaccessibility, and legal loopholes. Incorporating perspectives on technology, ethics, and human rights, the chapter offers an in depth account of today's reproductive practices that reflect both the perplexity of the modern world and the ongoing social inequalities that have persisted for centuries.

Keywords Assisted reproductive technology · Surrogacy · Designer babies · In vitro fertilization

6.1 Introduction

A global health concern that affects both men and women of reproductive age is infertility, which is becoming more common due in part to improved diagnosis and increased awareness. Assisted reproductive technology or ART is regarded as the most effective line of infertility treatment (Medenica et al., 2022). Medical interventions in ovulation, Fertilization, and the implantation of fertilized eggs into the uterus are all considered reproductive technology. To achieve Fertilization and implantation, eggs, sperm, and embryos are usually handled during ART treatments.

A. Shibi Anilkumar, R. Veerabathiran, *Bioethics and Disability*, SpringerBriefs in Modern Perspectives on Disability Research,
https://doi.org/10.1007/978-981-95-8197-9_6

The most popular ART process is IVF, which will be discussed primarily together with related methods like cryopreservation and intracytoplasmic sperm injection (ICSI) (Tenchov & Zhou, 2025). Reproductive technologies have benefited the most significant number of people when viewed from a utilitarian perspective (Nnachi et al., 2024).

Unfavorable short- and long-term perinatal outcomes are inconsistently linked to assisted reproductive technology (ART). The use of IVF in single females and same-sex couples, the "ownership" of gametes and embryos, intrafamilial gamete donation, the use of preimplantatory genetic testing, social egg freezing, commercialization, public funding, and the prioritization of IVF are just a few of the many intricate ethical, moral, and financial issues related to ART (Graham et al., 2023). There are questions about whether assisted reproductive technologies fully grant humans the respect that is owed to them from the moment of creation, their dignity, their sexuality, and the transmission of life. There are many ethical concerns about the essential values of sexuality, marriage, and parenthood, as well as potential societal and economic fallout from such reproductive treatments (Nnachi et al., 2024).

The instinct for parenthood is one of the strongest drives across all living beings. In many societies, infertility is viewed not as a medical condition alone but as a personal tragedy, often associated with failure to fulfil the expected biological role of reproduction. Historically, especially during the Common Era (c. 300–400 CE), a woman's value was tied first to her purity and later to her reproductive capacity. During the late medieval and early modern periods, various religious explanations for infertility coexisted. Although the changing concepts of gender equality influenced the views on procreation as necessary for the continuity of lineage and family, the different interpretations of infertility still prevailed. The scientific method was first applied to infertility when doctors began to look for its causes. This ultimately led to major revelations such as Spallanzani's early experiments in artificial insemination, which paved the way for the first successful attempt in a woman. The subsequent rapid development of assisted reproductive technologies (ART) has become one of the most significant medical achievements for addressing infertility worldwide (Sharma et al., 2018).

The reproductive justice movement, which was established by African American women in the United States in 1994, raised the concern that the foremost abortion rights advocates were too narrow in their approach. The right to terminate a pregnancy as the only reproductive right made the issue a matter of individual choice and deprived it of the broader reproductive needs. The limited scope completely neglected important issues like sterilization against one's will, access to IVF and surrogacy, prenatal and postnatal care, and the social conditions that must be met for raising children, like economic security, basic housing, stable food supply, and safety. Reproductive justice not only provided an intersectional framework but also recognized that various forms of oppression overlap. Thus, the grouping of people based on their social identities and membership determines the degree to which their reproductive lives are influenced by such oppression (McLaren & Rajat, 2025).

Numerous reproductive technologies are promoted as means of empowering people to take charge of their own destiny. Despite its unavoidable decline, people are assured that storing their eggs can guarantee long-term fertility. Reproductive technologies constitute an enormous and varied research area spanning multiple disciplines. The concept of reproductive justice, with its innovative approach to the whole, has a pivotal influence and a significant impact on every aspect of technology research. When problems arise, scholars must rely on it. According to the reproductive justice perspective, parents of children with special needs bear the burden of the lack of social support, not the disabled person. That argument is consistent with what academics and disability justice activists have written about the "politics of perfect babies" (Denbow, 2025).

There are new prospects for parents with disabilities at the nexus of legal parenting, technology, reproductive health, and disability rights. The rights of disabled parents have been infringed throughout history (Rothler, 2017). People with impairments may be unfairly subjected to paternalistic criticism, particularly criticism over their capacity to give consent for sexual activity or procreation. Too frequently, people with disabilities are portrayed as requiring special protection, including policies that limit their aspirations for close relationships and family life. People with disabilities may unfairly have less access to medically necessary reproductive care than people of the same age and sex due to assumptions such as these. Concerns regarding the risk of pregnancy when a person with a disability is involved are similarly misguided (Silvers et al., 2016).

Patients' disabilities may pose serious risks for future children, including the chance of life-threatening or significantly debilitating impairments. Some physicians object ethically to assisting pregnancy under such circumstances. They may also worry about indirect risks such as the impact of a parent's life-threatening condition on a child's welfare and the broader ethical issues surrounding decision-making and parental autonomy (Coleman, 2002). This chapter discusses assisted reproduction, surrogacy, and genetic selection. It looks at how these technologies interact with long-standing social, ethical, and disability-related issues concerning the politics of family formation, the experiences and rights of disabled people within these systems, and broader implications for reproductive autonomy.

6.2 Designer Babies and Disability

Designer babies are offspring of in vitro fertilization (IVF) embryos selected based on the presence or absence of specific genes. Preimplantation genetic diagnosis (PGD) is used to select "disease-free" embryos to prevent the majority of designer offspring from acquiring genetic abnormalities. By actively removing the illness gene from carrier embryos, scientists can produce designer babies using recently developed gene-editing technologies (Pang & Ho, 2016). Ethical tensions surrounding designer babies, illustrating how perceived benefits such as disease prevention and reproductive control simultaneously raise significant moral concerns are mentioned in Table 6.1.

Table 6.1 Benefits and ethical concerns of designer babies

Potential benefit	Description	Ethical counter-concern
Prevention of genetic disease	Avoids hereditary disorders	May devalue existing disabled lives (expressivist argument)
Reduced health burden	Eliminates costly or fatal diseases	Creates a genetic poor–rich divide based on affordability
Precision editing	CRISPR allows targeted gene modification	Unknown long-term risks; identity and psychological harm to the child
Enhanced traits	"Better" physical/cognitive traits	Leads to eugenics and genetic inequality
Greater reproductive control	Parents choose characteristics	Treats the child as a product rather than a person

Recent developments in reproductive medicine could eventually allow parents to alter their child's embryos. The assisted reproductive technologies (ART), like PGD and IVF, are practically the only option for the majority of infertile couples. One of the things PGD can do is determine the genetic problems that have the greatest impact on embryos before actual implantation. Through this process, it is assured that during the IVF fertility treatment, only those embryos that do not carry any genetic disorders will be transferred to the womb. Before implantation, those with the dangerous genes will be eliminated. These days, couples can avoid having a child with a hereditary illness by using PGD and IVF (Turriziani, 2014).

6.2.1 Development and Debate Around PGD

Preimplantation genetic diagnosis (PGD) was created to assist individuals or couples who are at risk of passing on severe hereditary diseases by performing in vitro fertilization (IVF) and implanting only embryos free of harmful mutations. As a process that allows the genetic testing and selection of embryos, PGD is still one of the most controversial innovations in the field of assisted reproduction. Its assumed capability to produce "designer babies" has given rise to both optimistic and fearful reactions about the existence of genetically modified or socially stratified groups of humans, which are the same visions that have been around since the 1930s (Löwy, 2018).

Genome editing, the reengineering method responsible for the birth of designer babies, is still a nascent technology and is far from widely available. Presently, the use of genome editing is predominantly confined to research, and it has been applied to a limited number of cases in the clinic, such as modifying the genomes of adult patients' cells to cure genetic diseases (Hisan & Romero, 2023).

6.2.2 Technologies Enabling Designer Babies

Techniques, including CRISPR-Cas9 and advances in genetic engineering, are enabling researchers to alter specific genes. Genetic modification can be used either to remove inherited disorders completely or to reduce the risk of developing certain illnesses. CRISPR-Cas9 alters DNA in a relatively fast, cheap, and precise way. Many academics are optimistic about these advances. Transhumanists' excessive faith in genetic technologies diverts our attention from improving the social determinants of health, including fetal health. They see genetic engineering as a "miracle cure" that can improve the human species and open new horizons (Knight, 2023).

6.2.3 Ethical Concerns

However, there are several moral issues related to designer babies. Some contend that it might result in a society where the "genetically privileged" and the "naturally born" are separated by genetic enhancement. The possibility of eugenics, in which some features are preferred over others and people who lack those traits are discriminated against or devalued, is a source of concern. Many nations have legal and regulatory systems in place that limit or prohibit genetic changes for nonmedical purposes, such as improving features or producing "designer babies," due to ethical concerns (Sanjay & Prasath, 2023).

The ethical dilemma of genetic editing for designer babies includes concerns that it dehumanizes children by treating them as customized items, along with the fear of increasing social inequality since these technologies might be accessible only to the rich, which are the two main issues. Moreover, critics also point out unintended negative impacts that might come with it, such as genetic uniformity, unidentified health risks, and alterations in societal values, alongside warnings of a possible return to an eugenic mindset that accepts only certain traits as better. Strong moral and theological concerns are raised by those who support reproductive autonomy and the freedom to select a child's characteristics. In general, it is yet unclear how genome editing will affect society and human development in future generations (Hisan & Romero, 2023).

Since the possibility of human genetic modification first surfaced, genetic enhancement has been discussed because it can alter the body in ways that are undetectable and permanent. The same ethical issues that underpin the traditional rejection of human improvement as a scientific objective would be raised by the use of human gene editing to prevent disease (Juengst et al., 2018).

6.2.4 *Disability, Social Marginalization, and Acceptability*

Disabled individuals are still among the disadvantaged population that are at risk of social marginalization. The meaningful involvement in societal activities is limited by individual-level marginalization, resulting in unemployment and, hence, poverty. Disability, social exclusion, and poverty are inextricably related; despite poverty alleviation efforts, disability reforms, advocacy, and disability acts, this disadvantaged group of disabled individuals nonetheless remains as victims (Sarkar & Parween, 2021).

In society, specific disabilities are more or less considered tolerable, sometimes eliciting sympathy or even approval, mostly because their causes are understood and accepted. On the other hand, disabilities related to fear, anxiety, depression, or confusion are not tolerated, regarded as signs of weakness, and thus concealed. The source and nature of the disability, as they are perceived, greatly influence its classification as acceptable or unacceptable. These social judgments become internalized, leading individuals with unacceptable disabilities to conceal them, avoid medical help, and continue working despite distress, not only to prevent financial loss but to protect their self-worth. Such individuals often wait for a "face-saving" reason before revealing their condition. This demonstrates how societal attitudes determine what counts as an acceptable versus an unacceptable disability (Hirschfeld & Behan, 1966).

Beyond the generational implications of genetic modification, concerns also arise regarding its impact on personal identity. Developing a stable sense of self is emotionally essential in childhood and adolescence, as it shapes resilience and character. If parents disclose that CRISPR or related technologies were used in designing their child's genetic makeup, it may profoundly influence how the child views themselves and their sense of who they were "supposed" to be. Using genetic engineering to obtain particular traits could also generate internal conflict, especially when the child compares themselves with peers who were naturally conceived. Some individuals might even experience guilt over the advantages enabled by technology, further complicating their self-understanding and emotional development (Page, 2024).

6.2.5 *Broader Criticisms of Designer Babies*

Designer babies have drawn much criticism for some reasons: they raise serious ethical issues, such as concerns about eugenics and unintended harm to future generations; they may reduce genetic diversity if many people choose the same preferred traits; they run the risk of increasing social inequality because such technologies would probably only be available to the wealthy; they involve unpredictable long-term effects on human health and society; and many people oppose them on moral or religious grounds (Hisan & Romero, 2023).

6.3 Ethics of Embryo Selection

At the end of the 1980s, preimplantation genetic diagnosis (PGD) was authorized for use by reproductive couples at risk of transmitting disorders associated with the X chromosome to their children. The genetic analysis was performed after aspirating one or more cells from an embryo produced through an in vitro fertilization (IVF) procedure. Preimplantation genetic diagnosis (PGD) has sparked widespread discussion worldwide and is often viewed as a contentious process. Although PGD couples are typically viable, they must undergo IVF treatment, which can be expensive and stressful (Sciorio et al., 2020).

Preimplantation genetic screening (PGS) and preimplantation genetic diagnosis (PGD) have historically been distinguished from one another. A case group with "hereditary disease in future parents" is described by PGD, whereas a case group with "suspected genetic disorders at the level of gametes and embryos" is described by PGS (Schmutzler, 2019). Additionally, PGT approaches are employed in clinical settings to prevent the inheritance of genes linked to diseases with delayed onset, diseases that can be cured, and an elevated but not absolute risk of disease. Additionally, it is used to select for specific genetic features and to detect sex embryos for family balancing (Imudia et al., 2016).

Preimplantation genetic testing for aneuploidy (PGT-A), formerly known as preimplantation genetic screening (PGS), is an IVF cycle in which embryos are biopsied and checked for chromosomal abnormalities before being placed into the uterus. Using fluorescence in situ hybridization (FISH) and a small number of chromosome-specific probes (usually 7–9 chromosomes), geneticists began attempting genetic screening at the cleavage-stage biopsy and single-cell analysis (Sciorio & Dattilo, 2020). Preimplantation genetic testing (PGT) for monogenic diseases (-M) can be used to maximize both IVF success and the quality of life for the future child when the patient or spouse has a known susceptibility to passing on genetic illnesses to their offspring (Amin et al., 2025).

6.3.1 Ethical Dilemmas: Severity, Selection Criteria, and Enhancement

Embryo selection methods generally offer little benefit to patients with average prognoses and can even harm those with poor prognoses. While PGD allows women or couples greater control over which embryos are transferred, it raises significant ethical concerns. One dilemma is whether the goal of achieving a "healthy baby" should include accepting embryos that test as carriers, children who will not experience the disease themselves but may transmit it to future generations. Rejecting carrier embryos risks extending PGD beyond the health of the immediate child and moving toward a slippery slope of progressively narrowing definitions of what counts as "acceptable." A second dilemma involves using PGD for sex selection for nonmedical, social reasons, which could contribute to a slippery slope toward broader nonmedical selection and rising parental expectations (Ehrich et al., 2007).

The most commonly used criterion for supporting the use of PGT to identify a genetic disease is severity. Public laws and regulations governing the use of PGT also frequently refer to disease severity. Their impact on pregnancy outcomes often justifies the detection of "severe" conditions through PGT, since some aneuploidies significantly increase the risk of miscarriage. Severity, therefore, guides which conditions should be screened to improve the likelihood of carrying a pregnancy to term. Overall, the more severe a disease is perceived to be, the greater the consensus on the Acceptability of using PGT (Gallois et al., 2025).

In the absence of unaffected embryos, PGT specialists believed that transferring a positive embryo was "an understandable choice," although opinions differed based on illness severity. Concerns included defining severity thresholds, potential societal financial strain, and parental attitudes toward a child whose genetic status is known from birth (Besser et al., 2019). Enhancement may be used not to give children "extra advantages," but to ensure they have the basic abilities needed for a good life, or to prevent their well-being from deteriorating in a changing or threatening world (Haqq, 2014).

6.3.2 *Embryo Selection in Practice: Quality of Life, Disability, and Professional Autonomy*

At the end of an IVF cycle, multiple viable embryos are often available for transfer, and current practice encourages single-embryo transfer to reduce the risks of multiple pregnancies without lowering success rates. Embryos are not considered harmed if transferred later or not transferred at all, so choices are guided by ethically acceptable reasons, such as the shortest time to pregnancy and preventing births likely to affect lifespan or quality of life severely (Polyakov et al., 2023).

Ethical issues arise in the case of selecting against disability, but it is a practice that both clinicians and ethicists generally accept. Fertility doctors, upon medical judgment and the principle of nonmaleficence, can use their professional freedom not to perform the transfer of embryos with considerable abnormalities. Nonetheless, the situation might interfere with the patient's right to make decisions about their treatment and thereby necessitate open and straightforward communication, as well as proper calculation of both the medical and the patient's value risks (Polyakov et al., 2023).

Preimplantation genetic diagnosis (PGD) is occasionally applied for Human Leukocyte Antigen (HLA) matching to generate a "savior sibling" and very rarely, to deliberately choose embryos with specific genetic disabilities. Among such embryos, those that tested positive for deafness or dwarfism have been picked for the implantation process. Proponents of this practice say that it is advantageous for a child to be raised in a family/community with which they identify (e.g., Deaf culture). Critics worry it may not serve the child's best interests and might set a precedent for selecting embryos with more severe disorders in the future (Damiano, 2011). Carrier couples' reproductive decision-making is also shaped by ethical considerations such as moral views on pregnancy termination or attitudes toward impairments. Preimplantation genetic testing (PGT) may exacerbate socioeconomic disparities (Zhang et al., 2022).

6.3.3 Societal Implications and the Expressivist Argument

The rising capacity to identify fetal disorders, even those slight ones, may result in a greater number of medically motivated abortions. Doctors and therapists often highlight the negative side of the disability rather than offering the patients a balanced perspective, thereby hindering the informed decision-making process. If properly informed, some parents might be ready to take a child with a disability. Increased testing may lower the number of specific genetic or chromosomal disabilities, but it also risks harming the general acceptance of hereditary disabilities and disability as such. When prenatal diagnostics are used mainly to prevent the birth of "disabled" babies, it may erode the moral worth of disabled individuals and decrease social understanding and support (Kaye, 2023). The expressivist argument holds that choosing not to implant an embryo likely to have a disability expresses a negative or morally problematic view toward existing disabled people, implying that they are inferior or less deserving of existence (Gyngell & Douglas, 2018).

6.4 Disabled People as Surrogates and Recipients

6.4.1 Surrogacy and Assisted Reproductive Technologies (ART)

One significant use of assisted reproductive technology is surrogacy, in which a woman bears a child on behalf of another couple. Surrogacy is a crucial fertility treatment. With the development of in vitro fertilization (IVF), women without uteruses, with uterine abnormalities that prevent pregnancy, with severe medical conditions, or with other pregnancy-related contraindications can now become mothers by using their own or a donor's embryo that is transferred to the gestational carrier's uterus (Patel et al., 2018).

In the framework of assisted reproductive technology (ART) involving third parties, surrogacy presents practical, ethical, and legal issues, such as the gestational carrier's intended transfer of the child she has carried, the intended parent's relationship to the child and to the gestational carrier, the establishment of legal parenthood, and the process's potential commercialization. One of the primary causes of cross-border reproductive care, where individuals in need of a particular procedure travel abroad to receive treatments unavailable in their home country, usually in a commercial context, is the legislative prohibition on surrogacy in many nations (Shenfield et al., 2010).

6.4.2 Reproductive Rights and Barriers Faced by People with Disabilities

Although society has ignored their sexuality and reproductive concerns, goals, and human rights, persons with disabilities have the same rights to sexual and reproductive desires and expectations as those without disabilities. Individuals with impairments are infantilized and perceived as asexual (or, in certain situations, hypersexual), incapable of procreating, and unsuitable parents or spouses. People with disabilities continue to face challenges regarding their sexual and reproductive health and rights (SRHR), with women with impairments facing particular challenges (Addlakha et al., 2017).

Persons with disabilities around the world confront continuous reproductive injustices due to the existing ableist laws, policies, and practices, which impede their sexual and reproductive health and rights. The factors encompass SRHR services that are culturally unsuitable and difficult to access, denial of the sexuality of the people involved, and systematic abuses like forced or coerced sterilization and contraception. Feminist disability activists observe that justice for disabled people is still ignored or neglected within the SRHR movements and donors' priorities, mainly due to the absence of women with disabilities and the people of diverse sexual orientation, gender identity and expression, and sex characteristics (SOGIESC) in leadership roles. Their concerns are thus not sufficiently raised or given priority in the global human rights movement (Adams, 2023).

Due to disability-related prejudice, fertility clinics are free to deny these women equitable access to medical care, and there is no legal recourse. Many women fall into a legal gap where they are not considered "disabled enough" under the Americans with Disabilities Act (ADA) to receive protection, yet still face discrimination in reproductive care due to stereotypes about disability. Allowing this loophole to continue devalues disabled lives, restricts reproductive choices for women with genetic conditions, and further intensifies ongoing debates about fetal personhood laws (Popham, 2025).

6.4.3 Barriers in Healthcare Access and Reproductive Care

Access to medical offices, examination tables, and diagnostic equipment for women with disabilities is often restricted by barriers in the built environment. The factors related to women's health and their mobility issues can interfere with or influence directly or indirectly pregnancy, miscarriage, postpartum depression, sexual health and function, and even the choice of contraceptive methods. Moreover, harmful disability misconceptions and wrong ideas about the needs and wants of this group of women may directly influence the availability, the level of treatment, and its quality (Kalpakjian et al., 2020).

Individuals with physical disabilities are still regarded more as patients in need of help rather than as persons who can support and look after others. The path to parenthood for individuals with physical disabilities is a very difficult one, both socially and physically. Social misconceptions and unkind attitudes, alongside issues like strength, safety, and access, are among these difficulties. Even though numerous studies have indicated that having a parent with a physical disability may not harm child development, the social stigma and negative attitudes toward parents with physical disabilities continue to exist. These parents utilized a variety of adaptive methods and tools to create strong and positive communication with their children, even though there were environmental and attitudinal challenges in parenting with a disability (Dunne & Ryan, 2024).

6.4.4 Parenting with Disabilities

Parenting in the general population is recognized as negatively affected by low socioeconomic status, unemployment, and social exclusion or isolation. Compared to other parents, mothers and fathers with intellectual disabilities may be more likely to face these challenges. For instance, research indicates that a variety of interventions, such as parent education programs, can help parents with intellectual disabilities deal with issues that may impair their capacity to parent successfully (Coren et al., 2010).

6.4.5 Translational Surrogacy and Disability-Related Violations

An increasing number of disability-related rights violations are emerging globally within transnational surrogacy. Across several countries, documented cases show that intended parents, in their pursuit of parenthood, exploit economically vulnerable women as surrogates and, in some instances, refuse to accept children born with disabilities. Children with disabilities are socially rejected, and surrogates are unfairly held accountable for events outside of their control (Pérez & Rincón, 2025).

Surrogate moms and intended parents from the minority world have been the main subjects of research. The intended parents who live in the majority of Indigenous contexts, like India, have received little attention. For example, minority countries were the primary topic of a recent special issue on "Reproductive Mobilities." Because of this narrow perspective, surrogate mothers have been denigrated as "wombs" or "victims," intended parents as "buyers," and offspring as "purchased or designer babies" (Shah et al., 2022).

6.4.6 *Legal Gaps and the Rights of Disabled Children Born Through Surrogacy*

The present legal atmosphere continues to be unfavorable toward third-party reproductive alternatives, like adoption and surrogacy, for gay couples, single parents, and disabled individuals. Even though commercial surrogacy is steadily increasing, there is still a lack of thorough research on the impact of these practices on people with disabilities. The comprehension of the rights and dignity of children with disabilities born through surrogacy is minimal. The neglect of the intersection of disability rights and surrogacy points to the necessity of research to explore the effects of surrogacy practices on the dignity and rights of disabled children, especially in light of the unfavorable historical legacy of eugenics (Pérez & Rincón, 2025).

6.5 Conclusion

The use of assisted reproductive technologies is still changing the human reproduction landscape, but, at the same time, their advantages and disadvantages are still distributed unevenly. The current chapter reveals the dual nature of the aforementioned technologies, which are, on the one hand, capable of preventing the transmission of serious diseases and facilitating the creation of families, but on the other hand, are still stigmatizing, hence, reinforcing the existing social inequalities, and, eventually, pushing persons with disabilities to the edge of society. Ethical discussions about severity, acceptability, enhancement, and parental autonomy continue to highlight the contradictory relations between medical objectives and social values. The cases of disabled persons as parents, patients, and even surrogates expose the limitations imposed on them by stigma, discriminatory policies, and healthcare systems, all of which act as barriers to absolute reproductive freedom. The chapter argues that the combination of disability rights, reproductive justice, and careful regulation is crucial to preventing technological advances from repeating historical exclusion. Thus, equitable reproductive futures can only be achieved by giving priority to dignity, informed choice, and the total inclusion of disabled persons in all reproductive decision-making and policymaking processes.

References

Adams, L. (2023). Forgotten by donors: A call to action by persons with disabilities to resource disability justice within sexual and reproductive health rights funding. *Sexual and Reproductive Health Matters, 31*(3), 2261688. https://doi.org/10.1080/26410397.2023.2261688

Addlakha, R., Price, J., & Heidari, S. (2017). Disability and sexuality: Claiming sexual and reproductive rights. *Reproductive Health Matters, 25*(50), 4–9. https://doi.org/10.1080/09688080.2017.1336375

Amin, N., Kteily, K., Deniz, S., Faghih, M., Karnis, M. F., Amin, S., & Neal, M. S. (2025). The ART of Embryo Selection: A Review of Methods to Rank the Most Competent Embryo (s) for Transfer to Optimize IVF Success. *Biomedicines, 13*(11), 2766. https://doi.org/10.3390/biomedicines13112766

Besser, A. G., Blakemore, J. K., Grifo, J. A., & Mounts, E. L. (2019). Transfer of embryos with positive results following preimplantation genetic testing for monogenic disorders (PGT-M): experience of two high-volume fertility clinics. *Journal of Assisted Reproduction and Genetics, 36*(9), 1949–1955. https://doi.org/10.1007/s10815-019-01538-2

Coleman, C. H. (2002). Conceiving harm: Disability discrimination in assisted reproductive technologies. *UcLA Law Review, 50*, 17. https://doi.org/10.2139/ssrn.311319

Coren, E., Hutchfield, J., Thomae, M., & Gustafsson, C. (2010). Parent training support for intellectually disabled parents. *Campbell Systematic Reviews, 6*(1), 1–60. https://doi.org/10.4073/csr.2010.3

Damiano, L. (2011). When parents can choose to have the "perfect" child: Why fertility clinics should be required to report preimplantation genetic diagnosis data. *Family Court Review, 49*(4), 846–859. https://doi.org/10.1111/j.1744-1617.2011.01418.x

Denbow, J. (2025). Reproductive technology in Dystopian Times. *Sex and Sexualities*, 3033371251329579. https://doi.org/10.1177/3033371251329579.

Dunne, A., & Ryan, C. (2024). Being a parent with a physical disability: A systematic review. *Rehabilitation Psychology*. https://doi.org/10.1037/rep0000590

Ehrich, K., Williams, C., Farsides, B., Sandall, J., & Scott, R. (2007). Choosing embryos: Ethical complexity and relational autonomy in staff accounts of PGD. *Sociology of Health and Illness, 29*(7), 1091–1106. https://doi.org/10.1111/j.1467-9566.2007.01021.x

Gallois, H., Ravitsky, V., Roy, M. C., & Laberge, A. M. (2025). Defining ethical criteria to guide the expanded use of Noninvasive Prenatal Screening (NIPS): Lessons about severity from preimplantation genetic testing. *European Journal of Human Genetics, 33*(2), 167–175. https://doi.org/10.1038/s41431-024-01714-8

Graham, M. E., Jelin, A., Hoon, A. H., Jr., Wilms Floet, A. M., Levey, E., & Graham, E. M. (2023). Assisted reproductive technology: Short-and long-term outcomes. *Developmental Medicine and Child Neurology, 65*(1), 38–49. https://doi.org/10.1111/dmcn.15332

Gyngell, C., & Douglas, T. (2018). Selecting against disability: The liberal eugenic challenge and the argument from cognitive diversity. *Journal of Applied Philosophy, 35*(2), 319–340.

Haqq, L. I. (2014). *Choosing children, choosing the future: Pursuing immortality through child selection*. Doctoral dissertation. University of California.

Hirschfeld, A. H., & Behan, R. C. (1966). The accident process: III. Disability: Acceptable and unacceptable. *JAMA, 197*(2), 85–89. https://doi.org/10.1001/jama.1966.03110020073025

Hisan, U. K., & Romero, C. B. (2023). Designer babies are no longer science fiction: What are the ethical considerations? *Bincang Sains dan Teknologi, 2*(03), 124–132. https://doi.org/10.56741/bst.v2i03.437

Imudia, A. N., Schickler, R., Plosker, S. M., & Mikhail, E. (2016). A prospective randomized single blinded study of office based evaluation of patients presenting with abnormal uterine bleeding using concurrent office hysteroscopy and endometrial biopsy. Does the order of the procedures matter?. *Fertility and Sterility, 106*(3), e43. https://doi.org/10.1016/j.fertnstert.2016.07.135

Juengst, E. T., Henderson, G. E., Walker, R. L., Conley, J. M., MacKay, D., Meagher, K. M., Saylor, K., Waltz, M., Kuczynski, K. J., & Cadigan, R. J. (2018). Is Enhancement the Price of Prevention in Human Gene Editing? CRISPR J, 1(6), 351–354. https://doi.org/10.1089/crispr.2018.0040

Kalpakjian, C. Z., Kreschmer, J. M., Slavin, M. D., Kisala, P. A., Quint, E. H., Chiaravalloti, N. D., et al. (2020). Reproductive health in women with physical disability: A conceptual framework for the development of new patient-reported outcome measures. *Journal of Women's Health, 29*(11), 1427–1436. https://doi.org/10.1089/jwh.2019.8174

Kaye, D. K. (2023). Addressing ethical issues related to prenatal diagnostic procedures. *Maternal Health, Neonatology and Perinatology, 9*(1), 1. https://doi.org/10.1186/s40748-023-00146-4

Knight, A. (2023). Gene editing technologies, utopianism, and disability politics. *The Journal of Philosophy of Disability*. https://doi.org/10.5840/jpd20237319

Löwy, I. (2018). *Tangled diagnoses: Prenatal testing, women, and risk*. University of Chicago Press.

McLaren, M., & Rajat, S. (2025). Biopolitics and reproductive injustice: The medicalization of reproduction and transition. *Revista Ideação, 1*(51), 59–81. https://doi.org/10.13102/ideac.v1i51.11836

Medenica, S., Zivanovic, D., Batkoska, L., Marinelli, S., Basile, G., Perino, A., et al. (2022). The future is coming: Artificial intelligence in the treatment of infertility could improve assisted reproduction outcomes—the value of regulatory frameworks. *Diagnostics, 12*(12), 2979. https://doi.org/10.3390/diagnostics12122979

Nnachi, A., Nwinya, C., & Ogoko, A. (2024). Ethical implications of artificial reproductive technologies. *African Journal of Politics and Administrative Studies, 17*(2), 82–100. https://doi.org/10.4314/ajpas.v17i2.5

Page, S. (2024). *Designer babies: A phenomenological study on genetic engineering*.

Pang, R. T., & Ho, P. C. (2016). Designer babies. *Obstetrics, Gynaecology and Reproductive Medicine, 26*(2), 59–60. https://doi.org/10.1016/j.ogrm.2015.11.011

Patel, N. H., Jadeja, Y. D., Bhadarka, H. K., Patel, M. N., Patel, N. H., & Sodagar, N. R. (2018). Insight into different aspects of surrogacy practices. *Journal of Human Reproductive Sciences, 11*(3), 212–218. https://doi.org/10.4103/jhrs.JHRS_138_17

Pérez, A. C., & Rincón, A. M. G. (2025). Surrogacy and disability: An overview of the legal, regulatory, and ethical issues. *Oñati Socio-Legal Series*. https://doi.org/10.35295/osls.iisl.2383.

Polyakov, A., Rozen, G., Gyngell, C., & Savulescu, J. (2023). Novel embryo selection strategies-finding the right balance. *Frontiers in reproductive health, 5*, 1287621. https://doi.org/10.3389/frph.2023.1287621

Popham, K. L. (2025). Embryos are not people, but disability is difference. *Columbia Law Review, 125*(1), 193–234.

Rothler, R. (2017). Disability rights, reproductive technology, and parenthood: Unrealised opportunities. *Reproductive Health Matters, 25*(50), 104–113. https://doi.org/10.1080/09688080.2017.1330105

Sanjay, S., & Prasath, N. H. (2023). Designer Babies: Revealing the ethical and social implications of genetic engineering in human embryos. *International Journal of Science and Research (IJSR), 12*(7), 688–693. https://doi.org/10.21275/SR23710130528

Sarkar, R., & Parween, S. (2021). Disability and exclusion: Social, education and employment perspectives. *Bhutan Journal of Research and Development, 10*(2). https://doi.org/10.17102/bjrd.rub.10.2.003

Schmutzler, A. G. (2019). Theory and practice of preimplantation genetic screening (PGS). *European Journal of Medical Genetics, 62*(8), 103670. https://doi.org/10.1016/j.ejmg.2019.103670

Sciorio, R., & Dattilo, M. (2020). PGT-A preimplantation genetic testing for aneuploidies and embryo selection in routine ART cycles: Time to step back? *Clinical Genetics, 98*(2), 107–115. https://doi.org/10.1111/cge.13732

Sciorio, R., Tramontano, L., & Catt, J. (2020). Preimplantation genetic diagnosis (PGD) and genetic testing for aneuploidy (PGT-A): Status and future challenges. *Gynecological Endocrinology, 36*(1), 6–11. https://doi.org/10.1080/09513590.2019.1641194

Shah, S., Ergler, C., & Hohmann-Marriott, B. (2022). The other side of the story: Intended parents' surrogacy journeys, stigma and relational reproductive justice. *Health & Place, 74*, 102769. https://doi.org/10.1016/j.healthplace.2022.102769

Sharma, R. S., Saxena, R., & Singh, R. (2018). Infertility & assisted reproduction: A historical & modern scientific perspective. *The Indian Journal of Medical Research, 148*(Suppl), S10–S14. https://doi.org/10.4103/ijmr.IJMR_636_18

Shenfield, F., De Mouzon, J., Pennings, G., Ferraretti, A. P., Nyboe Andersen, A., De Wert, G., … & ESHRE Taskforce on Cross Border Reproductive Care. (2010). Cross border reproductive care in six European countries. *Human Reproduction, 25*(6), 1361–1368. https://doi.org/10.1093/humrep/deq057

Silvers, A., Francis, L., & Badesch, B. (2016). Reproductive rights and access to reproductive services for women with disabilities. *AMA Journal of Ethics, 18*(4), 430–437.

Tenchov, R., & Zhou, Q. A. (2025). Assisted reproductive technology: A ray of hope for infertility. *ACS Omega*. https://doi.org/10.1021/acsomega.5c01643

Turriziani, J. V. (2014). Designer babies: The need for regulation on the quest for perfection.

Zhang, J., Pastore, L. M., Sarwana, M., Klein, S., Lobel, M., & Rubin, L. R. (2022). Ethical and moral perspectives of individuals who considered/used preimplantation (embryo) genetic testing. *Journal of Genetic Counseling, 31*(1), 176–187. https://doi.org/10.1002/jgc4.1471

Chapter 7
Disability Law, Global Bioethics, and Clinical Justice

Abstract The laws of disabilities, bioethics, and clinical governance are changing worldwide, and as a result, the perception of the rights and healthcare needs of the disabled people is changing across cultures. The current chapter depicts the interaction of disability law, international bioethics, and clinical justice frameworks as means of inclusion, protectors of autonomy, and healers of inequalities. The CRPD, the ADA, and the RPwD Act are some of the legal measures that have been taken in Europe, the USA, India, and South Africa to widen the scope of protection legally and, at the same time, reveal the remaining gaps in terms of access, equity, and clinical practices. On the other hand, the latest biotechnologies, genomic testing, and artificial intelligence (AI) not only expand the debate but also introduce new ethical concerns about discrimination, human diversity, and the future of disability. The WHO, UNESCO, and UNICEF are global institutions that establish governance models that promote dignity, rehabilitation, and inclusive education. By incorporating the principles of disability justice, ethical decision making is made easier, and health systems and technological advancement are ensured to be based on rights, socially responsive, and inclusive.

Keywords Clinical justice · Inclusive healthcare · Ethical governance · International policy frameworks · Bioethics committees

7.1 Introduction

People with disabilities (PwDs), in contrast to the common population, experience lower health outcomes, including early demise, higher morbidity, and functional limits. For instance, PwDs have elevated rates of chronic illnesses and communicable diseases, are more likely to die younger, and have functional limits (Gréaux et al., 2023). These health discrepancies are caused by many factors, including difficulties obtaining medical care, a higher prevalence of behaviors that compromise

A. Shibi Anilkumar, R. Veerabathiran, *Bioethics and Disability*, SpringerBriefs in Modern Perspectives on Disability Research,
https://doi.org/10.1007/978-981-95-8197-9_7

health, and a lower likelihood of receiving health promotion and disease precautionary services compared to individuals without disabilities (Ginis et al., 2021).

People with disabilities (PwDs) have historically received inadequate treatment in the healthcare system, including unequal access to clinical and preventative care as well as public health and wellness programs. Health disparities are also caused and made worse by healthcare professionals' poor diagnosis, treatment, and skills. An ultimate theory of a justice-oriented tactic is the medicalization of impairments, which shapes disability as a shortage and serves as a gatekeeper to sources, supplies, civil rights, and shelters. Healthcare for all patients can be improved by a social justice paradigm that emphasizes PwD as a crucial component of the medical process (Engelman et al., 2019).

Disability-inclusive healthcare requires a complex comprehension of social justice, equity, and their significance to health accessibility. While these tenets are gradually becoming the core of worldwide health debates, the majority of LMICs (low- and middle-income countries) are still not using them correctly (Azubuike et al., 2025). All opportunities for social and economic advancement are denied to disabled people. They are deprived of fundamental services, including occupation, medical services, and education. The persistence of ableism and disableism has been exposed by several ethics researchers who have invited and provided intimate, first-hand stories of disability at work in many national and cultural contexts. Scholars of management and organizations continue to pay much less attention to the consequences of ableism and disableism than to other stigmatized traits, even when it comes to issues like inclusion (Branzei et al., 2025).

Making urban infrastructure accessible is crucial for people with disabilities to be included and to be independent; the exclusion of disabled people continues to exist when the cities are not accessible. To understand this problem, it is necessary to examine government systems, disability rights movements, and broader social and economic contexts (Chou et al., 2024). Digital technology has led to a variety of digital urban lifestyles and online interactions between residents and their homes. This digital disparity impacts PwDs' city experiences (Kolotouchkina et al., 2022). Accessibility takes on new dimensions today, when technology plays a substantial role in our everyday lives. The extraordinary abilities of artificial intelligence (AI) are altering, and perhaps eradicating, conventional hurdles that have impacted PwDs (Almufareh et al., 2024).

The 2006 adoption of the UN Convention on the Rights of Persons with Disabilities (CRPD) led to a dramatic increase in the international legal debate surrounding the rights of disabled persons (Celik, 2017). Bioethical personalism recognizes the ethical requirement associated with the UN Convention on the Rights of Persons with Disabilities, which prescribes the principles of respect for intrinsic dignity and individual autonomy, including through the development of habilitative and rehabilitation programs aimed at the full inclusion and participation of disabled people in all areas of life (De Micco et al., 2024).

The disability rights movement and bioethics were still in their infancy 30 years ago. The observers of disability scholarship had made a note already in 1989 that issues regarding the life and death of the disabled were the primary concern of bioethical deliberations on disability. The evolution of molecular genetics resulted in the more widespread and uniform application of prenatal testing and the introduction of laws that facilitated the selective abortion of fetuses with disabilities diagnosed in the mother's womb. The defenders of the rights of PwD (DR) reacted quickly to these practices with criticism. Since few advocates had access to intellectual venues or possessed academic positions, DR arguments first received little attention. With time, their concerns, especially concerning the biased effects of prenatal policies, began to gather awareness in the subject of bioethics, revealing long-term disputes (Amundson & Tresky, 2007). This chapter studies approaches to develop inclusion, rights-based bioethics, and inspects how disability regulation interrelates with medical ethics, hospital systems, and international policy frameworks.

7.2 Comparative Frameworks

The disability rights frameworks of Europe, the United States, India, and South Africa are examined below. Table 7.1 summarizes the major legal foundations, policy mechanisms, achievements, and ongoing challenges shaping disability rights across four global regions.

7.2.1 Europe

The Amsterdam Treaty was the stepping stone for the European Union's disability rights progression for the most part. It was Article 13 of the European Community (EC) Treaty that empowered the EC to take action against discrimination based on disability. The 2000 Charter of Fundamental Rights was a significant document that strengthened these actions; it included Article 21, which stated that discrimination based on disability was not—and never would be—allowed, and Article 26, which was all about including disabled people in society (Ferri, 2020). Up till now, EU's disability policies were not treaty-based and were mainly relying on soft law. Still, the end of the 1990s was a milestone in the establishment of main strategies and the European Disability Strategy 2010–2020 (EDS), and the Strategy on the Rights of Persons with Disabilities 2021–2030, which were both significant actions by the Commission and promoted the European Union's disability law (Ferri, 2022).

The European Disability Strategy 2010–2020 was a prominent effort that aimed at promoting the inclusion of PwD not only in society but also in the single market by making the built environment, transportation, and ICT more accessible through legislation and standardization tools (Charitakis, 2013). Just before the CRPD, the EDS mandated the states to incorporate disability rights into all laws, policies, and standards (Davy et al., 2019). The European Union and all Member Nations ratified the CRPD through Council Decision 2010/48/EC, thereby incorporating the

Table 7.1 Comparative disability rights and policy frameworks across Europe, United States, India, and South Africa 2.2 *United States of America*

Region	Laws and treaties	Disability rights mechanisms	Major achievements	Challenges
Europe (EU)	• Amsterdam Treaty (Art. 13) • EU Charter of Fundamental Rights (Arts. 21 & 26) • CRPD ratified (2010/48/EC) • GDPR (2016)	• European Disability Strategy 2010–2020 and 2021–2030 • Anti-discrimination mandate • Data protection for vulnerable groups	• Mainstreaming disability rights across policies • Accessibility in transport, ICT, built environment • Strong privacy protections	• 25–65% of PwD struggle to access essential services • Healthcare access remains uneven • ECtHR remedies require lengthy domestic exhaustion
United States (USA)	• Rehabilitation Act (1973, Sec. 504) • Americans with Disabilities Act (1990, 2008 Amendments). • ACA (2010), Sec. 1557	• Anti-discrimination in employment, transport, digital access • Accommodation principle • Vocational rehabilitation (WIOA 2014)	• Civil rights model of disability • Broad legal protections in public and private sectors • Strong digital accessibility standards (Sec. 508, WCAG)	• Variability in enforcement • Healthcare and insurance gaps persist • Digital compliance inconsistent across states
India	• Constitution of India (Arts. 14, 15, 41) • PwD Act 1995; RPwD Act 2016 • National Trust Act 1999 • CRPD ratified (2007)	• Expanded disability categories (Article7–21) • Legal capacity, nondiscrimination, dignity • Niramaya Health Insurance Scheme • Education policies for inclusive schooling	• Strong constitutional and legislative framework • Protections for women/children, informed consent in research • Growing focus on health access and guardianship reform	• Limited awareness of schemes • Under-resourced health and rehabilitation systems • Uneven implementation across states
South Africa	• Constitution (1997, Section 9 equality clause) • Promotion of Equality & Prevention of Unfair Discrimination Act. • National Rehabilitation Policy (2000) • UNCRPD ratified	• Rights to nondiscrimination, access to health, transport, information • Emphasis on redressing historical inequalities • WHO Rehabilitation 2030 alignment	• Strong constitutional protections • Early adoption of CRPD • National Health Council supporting disability-inclusive policy	• Inadequate access to rehabilitation and assistive devices • Staff shortages, corruption, poor transport systems • PwD still excluded from essential health services

Convention into the EU legal system as a fundamental element. Thus, the EU secondary legislation in the field of the CRPD must be considered as a crucial factor (Ferri & Subic, 2023). The year 2016 marked another turning point when the General Data Protection Regulation (GDPR) came into force, ending a long and complex negotiation process that had been heavily influenced by Big Tech's lobbying and the lack of homogeneity of the privacy laws among the EU Member States (Mager, 2017). The GDPR not only repealed the 1995 Data Protection Directive but also established a more extensive and modernized structure for the entire European Union, including the digital health sector (Marelli et al., 2020).

The GDPR fortifies the right to data protection under the EU Charter and disallows the handling of "special categories of data," which encompass race, religion, and sexual orientation information, among others, except for very specific scenarios (Van Bekkum & Borgesius, 2023). It further underscores the importance of informational privacy to the dignity of children and vulnerable adults, echoing global instruments such as the UN Convention on the Rights of the Child, explicitly tying privacy to dignity in Article 88 (Piasecki & Chen, 2022).

Nonetheless, significant inequalities remain. One report from the EU Agency for Fundamental Rights revealed that between 25–65% of persons with disabilities in the Member States have a hard time getting at least one of the essential services like banking, public transport, and primary healthcare, and access to healthcare is the most difficult one (Groenewegen et al., 2021). At the level of the judiciary, the European Court of Human Rights (ECtHR) rules on cases involving human rights violations against disabled persons only after all domestic remedies have been exhausted. The court's decisions serve as normative references that shape the concepts of justice, human dignity, and fair access to healthcare for PwD (Skuban-Eiseler et al., 2023).

7.2.2 USA

In the past, around 35–43 million Americans lived with disabilities; thus, they formed one of the largest minority groups and communities. President Bush, in the presence of more than 3000 advocates, on July 26, 1990, sanctioned the Americans with Disabilities Act (ADA), thus marking a new period of optimism in the history of people with disabilities (Schall, 1998). This move was preceded by the Rehabilitation Act of 1973, which was the very first federal law to grant civil rights protection to people with disabilities. Among its major provisions was Section 504, which prohibited discrimination in all programs that received federal funding and later guided the formulation of the wider ADA (Schur et al., 2024).

The ADA of 1990 not only established design standards for accessible places but also prohibited discrimination based on disability. The 1991 ADA Standards for Accessible Design, which enumerate both technical and scoping requirements, were issued by the United States Department of Justice, and among the requirements were accessibility to transit vehicles and facilities (Aldousari et al., 2021).

The ADA, along with Section 504 and the ADA Amendment Act of 2008, which reemphasized the broad definition of disability, has been and remains the main framework for civil rights for people with disabilities (Iezzoni et al., 2022).

Unlike the 1964 Civil Rights Act, the ADA reframed disability as a social and environmental issue rather than a medical problem that limits individual capacity (Friedman & VanPuymbrouck, 2023). Americans with Disabilities Act (ADA) and its revisions issued in 2008 are the anti-discrimination laws of the United States, which are based on two concepts: overcoming both the attitudinal and structural barriers and the "accommodation principle" which mandates making reasonable changes in employment and public services (Blanck, 2022).

The Vocational Rehabilitation Program, initially established in the 1973 Rehabilitation Act, describes the pre-employment assistance for youngsters with disabilities. It was again redefined by the 2014 Workforce Innovation and Opportunity Act (WIOA) (Betz, 2023).The Affordable Care Act (ACA) of 2010 was totally executed in 2014, demonstrating a noteworthy shift in US health policy that aimed to reduce disparities and expand affordable healthcare (Vaitsiakhovich & Landes, 2023; Handel & Kolstad, 2015). Often known as Obamacare, it promoted the idea that healthcare is a right and sought to reduce the number of uninsured people (Hall & Lord, 2014). Section 1557 of the ACA strengthened protections against disability discrimination, while Section 5307 authorized federal funding for disability-competent care training (Iezzoni et al., 2022).

The 1973 Rehabilitation Act first created the Vocational Rehabilitation Program, and it fundamentally includes the provision of pre-employment support to disabled youths. The program was completely updated by the 2014 Workforce Innovation and Opportunity Act (WIOA) (Betz, 2023). The 2010 Affordable Care Act (ACA), which was fully implemented in 2014, was a major revolution in US health policy intended to eliminate inequalities and make it easier for the whole population to access affordable healthcare (Vaitsiakhovich & Landes, 2023; Handel & Kolstad, 2015). The Affordable Care Act, also known as Obamacare, promotes the idea that healthcare is a basic human right and aims to lower the uninsured rate (Hall & Lord, 2014). Section 1557 of the ACA also provided additional safeguards against disability discrimination, while Section 5307 allocated resources for training healthcare practitioners to provide appropriate care for PwD (Iezzoni et al., 2022).

To address digital accessibility, the State ADA Coordinator's Office ensured compliance with Section 508 and WCAG 2.1 (AA). Some of the practices they followed were easy navigation, content that a screen reader can read, good color contrast, text alternatives, structured headings, accessible forms, ARIA labels, and table markup. The new Final Rule by the US Department of Justice regarding web and mobile accessibility adds further requirements to these requirements as states work toward full ADA compliance (Georgia State ADA Coordinator's Office, 2020).

7.2.3 India

Every legal citizen of India is entitled to equal opportunities under the Indian Constitution, regardless of their physical, mental, or disability. People with disabilities are protected by the Indian Constitution's fundamental laws, legislation, and acts. The Indian Constitution grants its citizens equal rights to live with dignity and respect. It also applies to those with disabilities (Constitution of India, 1950) (Aneraye & Mousina, 2024).

The constitutional basis is reinforced by terms such as Article 14 (equality before the law), Article 15 (nondiscrimination), and Article 41 (right to employment, education, and aid), which defend the rights of PwD. Later, the PwD Act of 1995 and the RPwD (Rights of Persons with Disabilities) Act of 2016 paved the path for India's sanctioning of the UNCRPD in 2007, notably developing the disability rights framework, which was earlier welfare-based. The RPwD Act increased the disability regulations from seven to twenty-one sections. It ensured fundamental rights such as equality, dignity, nondiscrimination (Section 3), protections for women and children (Section 4), independence from cruelty and mandatory informed consent in research (Section 6), accessible voting (Section 11), accessible justice (Section 12), and equal legal capacity (Section 13) (Samyukta, 2025).

The National Trust Act (NTA), 1999, which is seen as historical legislation, further substantiated this framework by providing legal guardianship for disabled individuals after the age of 18. As per the National Institute for the Empowerment of Persons with Intellectual Disabilities (NIEPID), the National Trust for the Welfare of Persons with Autism, Cerebral Palsy, Mental Retardation, and Multiple Disabilities Act was enacted in 1999. Under this statute, a national-level administration was established that focused on the welfare not only of the mentioned persons with disabilities but also on related issues (Goswami Vernal, 2023).

To enhance access to healthcare, the National Trust introduced the Niramaya Health Insurance Scheme in 2008 to encourage PwD to obtain and use medical treatments. For a small premium, it provides yearly coverage of Rs. 1 lakh on a reimbursement basis. A distinctive feature is its coverage of outpatient services, including therapies (such as speech therapy or physiotherapy) and medications (such as levetiracetam and methylphenidate) that might not be offered in government hospitals near PwD's home. However, health professionals, people with disabilities, and caregivers are less aware of the Niramaya Insurance Scheme (Angothu et al., 2022).

The very beginning of the framework for disability policies in India has its roots in the Sargent Report of 1944, drafted by John Sargent, a British Chief Educational Advisor, and building on the 1936 CABE recommendation that provincial governments should not ignore the education of disabled children. The Kothari Commission further laid the groundwork by highlighting the inclusion of disabled children's education in general education and by suggesting facilities for four groups: the visually impaired, hearing-impaired people, the orthopedically challenged, and the mentally disabled (Aneraye & Mousina, 2024).

The 1968 National Education Policy was the one that eventually included all these suggestions. This policy aimed not only at the physically but also the mentally disabled children and thus gave rise to the movement for providing more educational facilities to the latter category of children. In 1986, the Rehabilitation Council of India was initiated as a society to monitor and standardize tutoring programs for professionals working with people with disabilities. Regarding the 1987 Mental Health Act, the emphasis was on human rights protections, and thus, the first step was taken in the right direction regarding the treatment of mentally ill patients. Section 81 of the Act disallowed all forms of indignity and cruelty during the treatment of persons with mental illness, while Section 94 made sure that their right to privacy in communications was not violated (Aneraye & Mousina, 2024).

7.2.4 South Africa

The Constitution of South Africa, prepared by the constitutionally elected assembly and coming into effect on February 4, 1997, was a significant achievement of the new democratic order. Among its significant aspects was the notion of equality. It was not only a core value of the Constitution, but, along with human dignity and freedom, the main principle for understanding the Bill of Rights and evaluating whether a restriction of rights was acceptable. Section 9, the equality clause, has three central pillars: (a) asserting that all persons are equally entitled to the protection of the law; (b) the ban of discernment and execution of unethical practices; and (c) the acknowledgment of taking measures to reimburse for difficulties brought about by discrimination. This section has played a critical role in many Constitutional Court judgments over the last 30 years (O'Regan, 2025).

One of the prime legal activities intended at repairing former racial inequalities in South Africa is the Promotion of Equality and Prevention of Unfair Discrimination Act, which is considered the most crucial, after the Constitution (Gutto, 2001). The Act's goal is to combat the long-standing effects of historical discrimination, which have contributed to the country's disparities in wealth and opportunities. The Act Section 4(2) pronounces to eliminate from agendas at all levels through the adoption of proactive measures about the inequalities and systemic discrimination, particularly concerning race, gender, and disability, and which traces back to colonialism, apartheid, and patriarchy (Kok, 2017).

African countries have a chief role in progressing the Convention on the Rights of PwD (CRPD). Deputies from Morocco, Sierra Leone, South Africa, Cameroon, Comoros, Mali, and Uganda enthusiastically collaborated with the Convention, whereas the authorities from Algeria, Kenya, and Tunisia made a notable contribution to its shaping. So far, the CRPD has been approved by 46 African countries and the Optional Protocol by 33, allowing entities to bring criticisms before the United Nations Committee on the Rights of PwD. These activities prove Africa's solid commitment to international disability rights standards (Veerabathiran & Thomas, 2025).

The World Health Organization (WHO), in 2017, issued Rehabilitation 2030: A Call for Action, which stressed the urgency of collaborative global action to make rehabilitation an essential part of health systems. For that, it aligns with the South African Constitution and the country's Disability Rights, which identify the rights of PwD to access high-quality healthcare and rehabilitation without barriers. Still, in South Africa, rehabilitation services are treated as inferior to curative medical services, which limits their recognition for enhancing the quality of life and economic growth. Thus, rehabilitation must become a national priority for strengthening interventions. In 1996, South Africa took its first step toward entrenching nondiscrimination based on disability in its Constitution, and the National Rehabilitation Policy of 2000 recognized the previous budget neglect while mandating provincial funding and human resources. The establishment of the National Health Council to steer health policy was another step in the right direction (Morris et al., 2021) through the National Health Act of 2003.

South Africa became one of the earliest countries to authorize the United Nations Convention on the Rights of PwD (UNCRPD) and thus provided protection for several rights, including the rights to information, transport, medical care, and living, among others. However, despite the legal framework, disabled persons are still being massively excluded from the health sector. They suffer from a lack of access to essential services, assistive devices, medication, and social support. The problems are even greater in lower-middle-income countries (LMICs), where impoverishment, lack of healthcare, poorly trained and equipped staff, transport that is inaccessible, corrupt practices, unstable governments, and negative perceptions of disability are the order of the day. In South Africa, these systemic issues elude people with disabilities, and inaccessible transport is still one of the significant barriers to timely access to healthcare (McKinney et al., 2021).

7.3 Global Disability Bioethics and Biotechnological Governance

7.3.1 Global Disability Bioethics

Global disability bioethics is an emerging approach that brings disability studies, human rights, and bioethics into conversation to examine how biomedical practices, biotechnologies, and health policies shape the lives, rights, and futures of disabled people across different global contexts. It places a great deal of importance on the personal experiences and understanding of disabled individuals, and perceives disability as a form of human diversity which has value rather than as an inadequacy, and interrogates how power, colonial histories, and global inequalities influence what counts as ethical in areas such as genetic screening, gene editing, assistive technologies, and public health priority-setting (Guidry-Grimes, 2022; Brocco, 2024; Ikecukwu, 2024).

At the dawn of the twenty-first century, biotechnological applications escalated significantly due to the rapid growth of molecular biology, which laid the groundwork for new scientific disciplines such as metabolomics, proteomics, and genomics (Martin et al., 2021). The pace of genetics and genomics in science and medicine is accelerating, and many technologies are being developed to pinpoint genetic variants that trigger disability-associated conditions. In genetic testing, only one gene is examined, whereas in genomic testing, multiple genes are analyzed simultaneously. Reproductive genetic testing is a technique that allows people to assess the likelihood of transmitting genetic conditions based on the carrier status. The aforementioned activities can significantly change the experiences of PwD. Besides, disabled persons along with their families have raised similar ELSI issues, such as reproductive freedom, loss of self-worth because of the use of technology that might preclude the birth of disabled children, and the impact on human diversity if the applications of genomic interventions may result in fewer births of people with genetic conditions (Vassos et al., 2024).

7.3.2 *Artificial intelligence (AI), Emerging Biotechnologies, and Implications for Disability*

There is substantial global evidence that the use of AI in healthcare is already making a significant impact, helping find solutions faster than usual by improving processes for early detection, diagnosis, treatment planning, and remote care delivery. Artificial intelligence (AI) is increasingly used in reproductive medicine to enhance selection and forecasting models for IVF, which are now more accurate than before (Gbagbo et al., 2024). The dramatic increase in the use of AI in human rights continues to be a legal challenge for the courts and legal doctrines. Artificial intelligence-based ruling already uncovered the presence of bias against some socially disadvantaged groups, like people with disabilities, and others who are discriminated against based on gender, skin color, or immigration status. United Nations Convention on the Rights of PwD (UNCRPD) does not explicitly mention AI, but it remains the most powerful global standard for safeguarding the rights of PwD. Its tenets provide a basic structure that can support the establishment of a rights-based control of AI systems through minimum safeguards (Bariffi, 2021).

In the past, the technologies for children with disabilities were primarily aimed at fixing some of their deficiencies. Nevertheless, rapid technological advances have changed treatment and prevention methods. Even though the availability of therapeutic and assistive tools has increased, there is still little evidence of their impact on children's functioning and quality of life. Despite many successful interventions, concerns persist that technological advances may also contribute to an apparent rise in childhood disabilities (Wise, 2012).

7.3.3 Global Governance: WHO, UNICEF, UNESCO, and CRPD Frameworks

The utilization of assistive technologies (AT) is increasing as the population ages, and the number of people with disabilities needing support grows. Still, there are major variances in access to AT within social groups and nations. Knowing how the world is addressing these gaps is essential. The United Nations, along with its partner organizations WHO and UNICEF, is leading the way in international cooperation through the UN. World Health Organization (WHO), the leading global health organization, aims to achieve the highest possible level of physical, mental, and social well-being for every person (Simeoni et al., 2023).

Patients' "disabilities" during treatment have been assessed and tracked using the WHO disability grading system (Brandsma & Van Brakel, 2003). The World Health Organization categorically promotes rehabilitation strengthening as one of the main components of the global health strategy (WHO Rehabilitation, 2020). Changes in health and demographic patterns are leading to a rise in the population of people who have permanent disabilities. Rehabilitation supports preventing temporary disabilities from becoming permanent, enhancing the effectiveness of other health services, and reducing overall healthcare costs (Negrini et al., 2020).

United Nations Educational, Scientific and Cultural Organization (UNESCO) advanced two themes through separate units: general education and special needs education. Over time, UNESCO helped lay the foundation for global educational inclusion, aiming to reach every child. In the mid-1990s, special needs education shifted toward the broader concept of inclusive education, and after 2000, it was merged into the Section for Combating Exclusion in Education. Since then, inclusive education has been aligned with the Education for All agenda. However, despite UNESCO's long-standing leadership in disability-related policymaking, recent versions of these programs have placed limited emphasis on the specific needs of learners with disabilities (Kiuppis, 2014). United Nations Educational, Scientific and Cultural Organization (UNESCO), as the global lead agency for education, can offer the long-term strategic direction needed to support WHO–UNICEF early childhood development (ECD) initiatives for children from birth to 5 years, as demonstrated in several high-income countries (Olusanya et al., 2022).

United Nations Educational, Scientific and Cultural Organization's (UNESCO's) UDBHR (Universal Declaration of Bioethics and Human Rights) acknowledges autonomy as one of its key bioethical principles, among others. Thus, in accordance with Article 5, humans have the right to make their own decisions, take responsibility for the effects of those decisions, and, concurrently, respect other people's independence (Rheeder, 2024). By 2005, UNESCO had already ratified the Universal Declaration on Bioethics and Human Rights, which comprises 15 essential ethical strategies along with guidelines on their scope, objectives, and use. These principles are intended to support ethical decision-making in global healthcare. The primary articles cover the topics of human dignity, benefit and harm, autonomy and consent, vulnerability, privacy, equality, nondiscrimination, cultural diversity, solidarity, social responsibility, sharing benefits, and, at the same time, the preservation of the environment and future generations. The ethical practice is further supported by additional basic principles, including nonmaleficence, beneficence, autonomy, justice, truthfulness, and trust (Jones-Bonofiglio et al., 2021).

United Nations Convention on the Rights of PwD (UNCRPD), approved in 2006, is considered the first human rights treaty made by disabled people for disabled people, and it is also regarded as a significant paradigm shift in disability rights. The Convention was an answer to the constant inequalities and discrimination that people with disabilities had to deal with, and it acknowledged that their rights, although equal to those of others, still needed particular safeguards to secure their complete involvement in society. The Convention's main principles are described in Article 3, and they uphold human worth, independence, dependence, equity, nondiscrimination, respect for diversity, participation and inclusion, accessibility, and recognition of the developing abilities of children with disabilities (Series, 2019; Smith et al., 2024). These different approaches combine to create a layered model for regulating biotechnology that would be scientifically reliable, respectful of human rights, and oriented toward justice.

7.3.4 Disability Justice, Social Inequality, and Ethical Futures

Disability justice and health justice are essential lenses for comprehending the nature of structural oppression. Disability rights, when seen through the traditional approach, deal mainly with individual suffering and legal compensation; on the contrary, disability justice takes a long view of systemic discrimination and the experiences of the most marginalized, especially those who are disabled and of color, or LGBTQ+. It denounces the discrimination based on one issue only and considers ableism as the chief contributor to disability inequality. The principles of the disability justice framework encompass ten concepts including the following: intersectionality, the predominance of affected people in leadership, solidarity across movements and disabilities, interdependence, collective access, funding sustainability, and the liberation of the whole community (Harris, 2022).

Some disability communities continue to have legitimate concerns that discoveries in genetics and genomics would be utilized to eradicate disability differences from the community, hence perpetuating the stigma associated with disabilities. These facts show how important it is to deliberately improve the views and values of individuals with disabilities in genetics and genomics. A collection of analytical tools for thinking through how to achieve this objective is provided by disability justice (Mintz et al., 2024).

New biotechnologies can greatly enhance human well-being, but also risk deepening existing inequalities. Claims of justice in access to healthcare goods and services derive their legitimacy from the essential role these resources play in supporting human well-being, which remains a core concern of social justice (Powers & Faden, 2006). In the current world, rapid developments in biomolecular science and genetic technologies present new opportunities but also raise significant ethical questions about how these inventions may reform citizenship and human dignity. As we move into a biotechnological future, it is critical to reminisce about past prejudices toward people with disabilities and ensure that developing technologies promote inclusion rather than repeat harmful practices (Clapton, 2003).

Biotechnology's evolutionary power makes ethical inspection inevitable, which in turn demands global governance frameworks that can control evolving life sciences in line with human rights standards; these same agendas progressively draw on disability rights such as the UNCRPD and disability-justice critiques of innovation, ensuring that decisions about biotechnological futures are shaped not only by hypothetical principles but also by the lived experiences, actions, and collective struggles of PwDs for impartiality and structural change (Trump et al., 2023; Benatar, 2002).

7.4 Inclusive Clinical Practice, Ethics Committees, and Advocacy in Healthcare

People with disabilities (PwDs) remain the least hired group, with an unemployment rate more than two-fold that of their nondisabled colleagues. The underrepresentation of PwDs in the labor market indicates a considerable problem of lost opportunity, as the hiring of disabled individuals has been linked to increased productivity both at the individual and organizational levels. One potential way to increase employment rates for PwDs is to raise companies' knowledge of the contributions PwDs can make to the workforce, particularly in companies that maintain a disability-inclusive mindset. The paradigm of a disability-friendly company is one that does not just act upon the regulations implied by anti-discrimination laws but rather tries to create a holistic and "disability-inclusive" atmosphere in every aspect of its operations, through its policies, and mindset. These companies deliberately work on spotting and removing obstacles like discrimination in communication, building, and the company's attitude itself that hinder PwDs from their full participation in the workplace. The case study eventually concludes that the disability-inclusive attitude, which sees great economic value, provides necessary support, and keeps in its ranks people with disabilities, can be brought about through frequent, company-wide talks and making sense of the situation together (Ochrach et al., 2022).

Inclusive employment, strong ethical oversight, reported incidents of discrimination, and workforce shortages at the system level determine whether the clinical practice is truly inclusive or just a slogan. Research indicates that employing people with disabilities and from various backgrounds leads to improved culturally competent care and patient outcomes, but this happens only when there are evident accountability systems and anti-discrimination provisions that shield the staff from reporting damages. Investigations into healthcare institutions reveal that the routine encounters of discrimination suffered by underrepresented physicians point out the exact places where the policies do not work in practice, thereby leading to burnout, high turnover, and inferior care. On the one hand, the structural workforce gaps consisting of insufficient workforce, inadequate accommodations, and continually minoritized and disabled professional nonrepresentation, among other factors, affect the time, flexibility, and relational work necessary for equity-oriented and person-centered care, and this means that even the best-designed inclusive practice guidelines cannot be fully implemented without simultaneous reforms in employment, supervision, and workforce planning (Naghipour et al., 2025; Marjadi et al., 2023; Bodicoat et al., 2021).

National Ethics or Bioethics Committees at the country level have the potential to improve the human rights situation of individuals with anomalies significantly. They are allowed to operate in accordance with internationally accepted rules, such as the World Medical Association's Declaration of Helsinki and the CIOMS International Ethical Guidelines for Health-Related Research Involving Humans. These instructions and guidelines enforce heavy obligations on scientists, especially those working with contributors who are vulnerable, and underscore that the research conducted must be of scientific and societal value, while guaranteeing the well-being, rights, and welfares of all humans are protected. The committees can notify society about the adverse impact of social barriers and the absence of support services on individuals with abnormalities, and promote gradual changes in approaches toward diversity in the human race and the acceptance and respect of the same as a norm. Moreover, it is their responsibility to provide references that are exact to the context to ease barriers, elevate support services, and enable people with impairments to fully involve in their community and society (Aarons, 2020).

Children with medical complexity (CMC), those with multimorbidity and physical restrictions, often experience disability-based discrimination in healthcare, whether through interpersonal interactions, institutional practices, or system-level policies. A study by Ames et al. (2024) explored parents' perceptions of this discrimination. Many parents reported a loss of trust in healthcare providers due to discriminatory experiences. They also described a heightened caregiving burden, including "forced advocacy," where they felt compelled to push for appropriate care for their child. One mother (parent 20) illustrated this constant need to advocate for her child with a disability. Discrimination against CMC also affected parents' own well-being, and those from minoritized racial or ethnic groups noted that racism and ethnicity-based bias were compounded by disability discrimination within the healthcare system (Ames et al., 2024).

The healthcare worker training concerning disability has some promising practices, but new methods may have to be applied in recruiting and hiring to cover a larger number of people with disabilities in the healthcare workforce. The majority of the good practice examples are directed toward improving rehabilitation, use of assistive technologies, and specialist services; nevertheless, there are still large deficits in leadership, financing, and affordability areas (Kuper et al., 2024).

7.5 Conclusion

The combination of disability law and global bioethics reveals that substantial inclusion, not just legal protection, requires a comprehensive transformation of structures. People with disabilities, despite international and national legal systems being helpful, still confront obstacles of a systemic nature in the areas of healthcare, clinical decision-making, digital environments, and biotechnological futures. Thus, ethical governance must place disability justice, intersectional analysis, and the lives of the most impacted at the top of the list of priorities. Rehabilitation, inclusive workforce policies, ethical oversight, and accessible technologies are to be enhanced to implement rights in practice. With the development of AI, genomics, and other biotechnologies, their use should be based on the values of freedom, respect, equality, and the good of the community. A rights-based, international-cooperation approach grounded in CRPD principles and aligned with WHO, UNESCO, and UNICEF frameworks offers the strongest path to creating health systems that are equitable, inclusive, and clinically just for all.

References

Aarons, D. E. (2020). The disability-rights perspective within the bioethics agenda. *Nursing Ethics, 27*(4), 1056–1065. https://doi.org/10.1177/0969733020906599

Aldousari, A., Alghamdi, A., & Alwadei, H. (2021). The 1991 Americans with Disabilities Act (ADA) standards for accessible design. *American Research Journal of Humanities and Social Sciences, 4*, 59–62.

Almufareh, M. F., Kausar, S., Humayun, M., & Tehsin, S. (2024). A conceptual model for inclusive technology: Advancing disability inclusion through artificial intelligence. *Journal of Disability Research, 3*(1), 20230060. https://doi.org/10.57197/JDR-2023-0060

Ames, S. G., Delaney, R. K., Delgado-Corcoran, C., Houtrow, A. J., Alvey, J., Watt, M. H., & Murphy, N. (2024). Impact of disability-based discrimination in healthcare on parents of children with medical complexity. *Developmental Medicine and Child Neurology, 66*(9), 1226–1233. https://doi.org/10.1111/dmcn.15870

Amundson, R., & Tresky, S. (2007). On a bioethical challenge to disability rights. *The Journal of Medicine and Philosophy, 32*(6), 541–561. https://doi.org/10.1080/03605310701680924

Aneraye, A. V., & Mousina, S. (2024). Development of Indian legislation for persons with disabilities since independence. *European Journal of Special Education Research, 10*(6). https://doi.org/10.46827/ejse.v10i6.5564

Angothu, H., Ajmera, S., Thanapal, S., Reddy, K. S., Jagannathan, A., Muliyala, K. P., & Thirthalli, J. (2022). Poor enrollment of persons with disabilities in niramaya health insurance scheme over a decade under the Indian national trust. *Indian Journal of Social Psychiatry, 38*(3), 297–300.

Azubuike, P. C., Akinreni, T., Adai, S. G., Ogbonna, C. K., Abba, M. E., Udofia, M. D., Odo, O. J., Nwadiche, M., & Imo, U. F. (2025). Equity and social justice perspectives on disability inclusion in healthcare services in Nigeria. *Communications Medicine, 5*(1), 371. https://doi.org/10.1038/s43856-025-01070-8

Bariffi, F. J. (2021). Artificial intelligence, human rights and disability. *Pensar-Revista de Ciências Jurídicas, 26*(2) https://periodicos.unifor.br/rpen/article/view/12704

Benatar, S. R. (2002). Human rights in the biotechnology era 1. *BMC International Health and Human Rights, 2*(1), 3. https://doi.org/10.1186/1472-698x-2-3

Betz, C. L. (2023). Health care transition planning for adolescents and emerging adults with intellectual disabilities and developmental disabilities: Distinctions and challenges. *Journal for Specialists in Pediatric Nursing, 28*(3), e12415. https://doi.org/10.1111/jspn.12415

Blanck, P. (2022). Disability-inclusive employment, cancer survivorship, and the Americans with Disabilities Act. *Journal of Cancer Survivorship, 16*, 142–151. https://doi.org/10.1007/s11764-021-01141-4

Bodicoat, D. H., Routen, A. C., Willis, A., Ekezie, W., Gillies, C., Lawson, C., Yates, T., Zaccardi, F., Davies, M. J., & Khunti, K. (2021). Promoting inclusion in clinical trials-a rapid review of the literature and recommendations for action. *Trials, 22*(1), 880. https://doi.org/10.1186/s13063-021-05849-7

Brandsma, J. W., & Van Brakel, W. H. (2003). WHO disability grading: Operational definitions. *Leprosy Review, 74*(4), 366–373. https://doi.org/10.47276/lr.74.4.366

Branzei, O., Zeyen, A., Bruyère, S. M., et al. (2025). Ethical dilemmas of disability inclusion—emerging ethics of activism and care: Introduction to the special issue. *Journal of Business Ethics, 201*, 781–795. https://doi.org/10.1007/s10551-025-06167-7

Brocco, G. (2024). Theories and practices of disability from the Global South: A critical anthropological perspective. *Frontiers in Health Services, 4*, 1261091. https://doi.org/10.3389/frhs.2024.1261091

Celik, E. (2017). The role of CRPD in rethinking the subject of human rights. *The International Journal of Human Rights, 21*(7), 933–955. https://doi.org/10.1080/13642987.2017.1313236

Charitakis, S. (2013). An introduction to the disability strategy 2010-2020, with a focus on accessibility. *Ars aequi, 1*, 28–35.

Chou, Y. C., Uwano, T., Chen, B. W., Sarai, K., Nguyen, L. D., Chou, C. J., et al. (2024). Assessing disability rights in four Asian countries: The perspectives of disabled people on physical, attitudinal and cultural barriers. *Political Geography, 108*, 103027. https://doi.org/10.1016/j.polgeo.2023.103027

Clapton, J. (2003). Ethics of Re-membering and remembering: Considering disability and biotechnology. *New Zealand Bioethics Journal, 21*.

Davy, L., Fisher, K., Wehbe, A., Purcal, C., Robinson, S., Kayess, R., & Santos, D. (2019). *Review of implementation of the National Disability Strategy 2010-2020*. https://doi.org/10.26190/5c7494b61edc4

De Micco, F., Tambone, V., Frati, P., Cingolani, M., & Scendoni, R. (2024). Disability 4.0: Bioethical considerations on the use of embodied artificial intelligence. *Frontiers in Medicine, 11*, 1437280. https://doi.org/10.3389/fmed.2024.1437280

Engelman, A., Valderama-Wallace, C., & Nouredini, S. (2019). State of the profession: The landscape of disability justice, health inequities, and access for patients with disabilities. *Advances in Nursing Science, 42*(3), 231–242. https://doi.org/10.1097/ANS.0000000000000261

Ferri, D. (2020). The unorthodox relationship between the EU Charter of Fundamental Rights, the UN Convention on the Rights of Persons with Disabilities and secondary rights in the Court of Justice case law on disability discrimination. *European Constitutional Law Review, 16*(2), 275–305. https://doi.org/10.1017/S1574019620000164

Ferri, D. (2022). The role of soft law in advancing the rights of persons with disabilities in the EU: A 'hybridity'approach to EU disability law. *European Law Journal, 28*(4–6), 134–153. https://doi.org/10.1111/eulj.12454

Ferri, D., & Subic, N. (2023). The European Union: Federal trends in disability rights. In D. Ferri, F. Palermo, G. Martinico, *Federalism and the Rights of Persons with Disabilities*. Hart Publishing (2023). https://www.bloomsbury.com/uk/federalism-and-the-rights-of-persons-with-disabilities-97815099624

Friedman, C., & VanPuymbrouck, L. (2023). Support for the Americans with Disabilities Act among nondisabled people. *Journal of Disability Policy Studies, 34*(3), 199–210. https://doi.org/10.1177/10442073211023175

Gbagbo, F. Y., Ameyaw, E. K., & Yaya, S. (2024). Artificial intelligence and sexual reproductive health and rights: A technological leap towards achieving sustainable development goal target 3.7. *Reproductive Health, 21*(1), 196. https://doi.org/10.1186/s12978-024-01924-9

Georgia State ADA Coordinator's Office. (2020). *Mission, purpose and website accessibility*. Available online: https://ada.georgia.gov/about-us/mission-purpose-and-website-accessibility

Ginis, K. A. M., van der Ploeg, H. P., Foster, C., Lai, B., McBride, C. B., Ng, K., et al. (2021). Participation of people living with disabilities in physical activity: A global perspective. *The Lancet, 398*(10298), 443–455. https://doi.org/10.1016/S0140-6736(21)01164-8

Goswami Vernal, T. (2023). Rights of persons with disabilities in India: Provisions, promises and reality. *BAU Journal-Society, Culture and Human Behavior, 4*(2), 4. https://doi.org/10.54729/2789-8296.1069

Gréaux, M., Moro, M. F., Kamenov, K., Russell, A. M., Barrett, D., & Cieza, A. (2023). Health equity for persons with disabilities: A global scoping review on barriers and interventions in healthcare services. *International Journal for Equity in Health, 22*(1), 236. https://doi.org/10.1186/s12939-023-02035-w

Groenewegen, P. P., Kroneman, M., & Spreeuwenberg, P. (2021). Physical accessibility of primary care facilities for people with disabilities: A cross-sectional survey in 31 countries. *BMC Health Services Research, 21*(1), 107. https://doi.org/10.1186/s12913-021-06120-0

Guidry-Grimes, L. (2022). Disability bioethics and the commitment to equality. *Theoretical Medicine and Bioethics, 43*(4), 209–220. https://doi.org/10.1007/s11017-022-09575-2

Gutto, S. (2001). *Equality and non-discrimination in South Africa: The political economy of law and law making*. New Africa Books.

Hall, M. A., & Lord, R. (2014). Obamacare: What the Affordable Care Act means for patients and physicians. *BMJ, 349*. https://doi.org/10.1136/bmj.g5376

Handel, B. R., & Kolstad, J. T. (2015). Health insurance for "humans": Information frictions, plan choice, and consumer welfare. *American Economic Review, 105*(8), 2449–2500. https://doi.org/10.1257/aer.20131126

Harris, J. E. (2022). Locating disability within a health justice framework. *Journal of Law, Medicine and Ethics, 50*(4), 663–673. https://doi.org/10.1017/jme.2023.6

Iezzoni, L. I., Rao, S. R., Ressalam, J., Bolcic-Jankovic, D., Agaronnik, N. D., Lagu, T., et al. (2022). US physicians' knowledge about the Americans with disabilities act and accommodation of patients with disability: Study examines what physicians know about the Americans With Disabilities Act and what is done to accommodate patients with a disability. *Health Affairs, 41*(1), 96–104. https://doi.org/10.1377/hlthaff.2021.01136

Ikecukwu, O. M. (2024). The intersection of bioethics and disability right. *Mexican Bioethics Review ICSA, 6*(11), 12–16. https://doi.org/10.29057/mbr.v6i11.13050

Jones-Bonofiglio, K., Wagoro, M. C. A., Arulappan, J., Brown, A., Cassells, K., Cooper, V., et al. (2021). *Journal of Nursing and Health Care Ethics Related to the COVID-19 Pandemic: Policy Guideline Recommendations Based on UNESCO's Universal Declaration of Bioethics and Human Rights*.

Kiuppis, F. (2014). Why (not) associate the principle of inclusion with disability? Tracing connections from the start of the 'Salamanca Process'. *International Journal of Inclusive Education, 18*(7), 746–761. https://doi.org/10.1080/13603116.2013.826289

Kok, A. (2017). The promotion of equality and prevention of unfair discrimination Act 4 of 2000: How to balance religious freedom and other human rights in the higher education sphere. *South African Journal of Higher Education, 31*(6), 25–44. https://doi.org/10.28535/31-6-1640

Kolotouchkina, O., Barroso, C. L., & Sánchez, J. L. M. (2022). Smart cities, the digital divide, and people with disabilities. *Cities, 123*, 103613. https://doi.org/10.1016/j.cities.2022.103613

Kuper, H., Gatta, D. R., Rotenberg, S., Banks, L. M., Smythe, T., & Heydt, P. (2024). Building disability-inclusive health systems. *The Lancet Public Health, 9*(5), e316–e325. https://doi.org/10.1016/S2468-2667(24)00042-2

Mager, A. (2017). Search engine imaginary: Visions and values in the co-production of search technology and Europe. *Social Studies of Science, 47*(2), 240–262. https://doi.org/10.1177/0306312716671433

Marelli, L., Lievevrouw, E., & Van Hoyweghen, I. (2020). Fit for purpose? The GDPR and the governance of European digital health. *Policy Studies, 41*(5), 447–467. https://doi.org/10.1080/01442872.2020.1724929

Marjadi, B., Flavel, J., Baker, K., Glenister, K., Morns, M., Triantafyllou, M., Strauss, P., Wolff, B., Procter, A. M., Mengesha, Z., Walsberger, S., Qiao, X., & Gardiner, P. A. (2023). Twelve tips for inclusive practice in healthcare settings. *International Journal of Environmental Research and Public Health, 20*(5), 4657. https://doi.org/10.3390/ijerph20054657

Martin, D. K., Vicente, O., Beccari, T., Kellermayer, M., Koller, M., Lal, R., et al. (2021). A brief overview of global biotechnology. *Biotechnology and Biotechnological Equipment, 35*(sup1), S5–S14. https://doi.org/10.1080/13102818.2021.1878933

McKinney, E. L., McKinney, V., & Swartz, L. (2021). Access to healthcare for people with disabilities in South Africa: Bad at any time, worse during COVID-19? *South African family practice : official journal of the South African Academy of Family Practice/Primary Care, 63*(1), e1–e5. https://doi.org/10.4102/safp.v63i1.5226

Mintz, K. T., Stramondo, J. A., & Tabor, H. K. (2024). Nothing about us without us in precision medicine: A call to reframe disability difference in genetics and genomics. *Hastings Center Report, 54*, S41–S48. https://doi.org/10.1002/hast.4928

Morris, L. D., Grimmer, K. A., Twizeyemariya, A., Coetzee, M., Leibbrandt, D. C., & Louw, Q. A. (2021). Health system challenges affecting rehabilitation services in South Africa. *Disability and Rehabilitation, 43*(6), 877–883. https://doi.org/10.1080/09638288.2019.1641851

Naghipour, A., Becher, E., Gemander, M., & Oertelt-Prigione, S. (2025). Designing clinical practice guidelines for equitable, inclusive, and contextualised care. *BMJ, 391*. https://doi.org/10.1136/bmj-2025-085684

Negrini, S., Kiekens, C., Heinemann, A. W., Ozcakar, L., & Frontera, W. R. (2020). Prioritising people with disabilities implies furthering rehabilitation. *The Lancet, 395*(10218), 111–111.

O'Regan, C. (2025). The long legacy of apartheid geography and the reach of the South African Constitution's Equality Clause. *German Law Journal, 26*(2), 198–212. https://doi.org/10.1017/glj.2025.14

Ochrach, C., Thomas, K., Phillips, B., Mpofu, N., Tansey, T., & Castillo, S. (2022). Case study on the effects of a disability inclusive mindset in a large biotechnology company. *Journal of Work-Applied Management, 14*(1), 113–125. https://doi.org/10.1108/JWAM-06-2021-0045

Olusanya, B. O., Halpern, R., Cheung, V. G., Nair, M. K. C., Boo, N. Y., Hadders-Algra, M., & Global Research on Developmental Disabilities Collaborators (GRDDC). (2022). Disability in children: A global problem needing a well-coordinated global action. *BMJ Paediatrics Open, 6*(1), e001397. https://doi.org/10.1136/bmjpo-2021-001397

Piasecki, S., & Chen, J. (2022). Complying with the GDPR when vulnerable people use smart devices. *International Data Privacy Law, 12*(2), 113–131. https://doi.org/10.1093/idpl/ipac001

Powers, M., & Faden, R. R. (2006). *Social justice: The moral foundations of public health and health policy*. Oxford University Press.

Rheeder, A. L. (2024). The ethical assessment of the stay-at-home order in South Africa in light of the universal declaration of bioethics and human rights (UNESCO). *Bioethical Inquiry, 21*, 229–237. https://doi.org/10.1007/s11673-023-10304-0

Samyukta, P. M. (2025). *Legal frameworks and accessibility in public spaces: Evaluating the implementation of disability rights in India*.

Schall, C. M. (1998). The Americans with Disabilities Act–are we keeping our promise? An analysis of the effect of the ADA on the employment of persons with disabilities. *Journal of Vocational Rehabilitation, 10*(3), 191–203. https://doi.org/10.3233/JVR-1998-10303

Schur, L., Ameri, M., Kruse, D., et al. (2024). Introduction to special issue: The new frontier of disability employment on the 50th anniversary of the rehabilitation act. *Journal of Occupational Rehabilitation, 34*, 279–282. https://doi.org/10.1007/s10926-024-10208-9

Series, L. (2019). Disability and human rights. *Routledge handbook of disability studies*.

Simeoni, R., Pirrera, A., Meli, P., & Giansanti, D. (2023, November). Promoting universal equitable accessibility: an overview on the impact of assistive technology in the UN, UNICEF, and WHO Web Portals. In *Healthcare* (Vol. 11, No. 21, p. 2904). MDPI. https://doi.org/10.3390/healthcare11212904.

Skuban-Eiseler, T., Orzechowski, M., & Steger, F. (2023). Access to healthcare for disabled individuals: An analysis of judgments of the European Court of Human Rights from an ethical perspective. *Frontiers in Public Health, 10*, 1015401. https://doi.org/10.3389/fpubh.2022.1015401

Smith, E. M., Huff, S., Wescott, H., Daniel, R., Ebuenyi, I. D., O'Donnell, J., et al. (2024). Assistive technologies are central to the realization of the Convention on the Rights of Persons with Disabilities. *Disability and Rehabilitation: Assistive Technology, 19*(2), 486–491. https://doi.org/10.1080/17483107.2022.2099987

Trump, B., Cummings, C., Klasa, K., Galaitsi, S., & Linkov, I. (2023). Governing biotechnology to provide safety and security and address ethical, legal, and social implications. *Frontiers in Genetics, 13*, 1052371. https://doi.org/10.3389/fgene.2022.1052371

Vaitsiakhovich, N., & Landes, S. D. (2023). The association between the Patient Protection and Affordable Care Act and healthcare affordability among US adults with intellectual disability. *Journal of Intellectual Disability Research, 67*(12), 1270–1290. https://doi.org/10.1111/jir.13037

Van Bekkum, M., & Borgesius, F. Z. (2023). Using sensitive data to prevent discrimination by artificial intelligence: Does the GDPR need a new exception? *Computer Law and Security Review, 48*, 105770. https://doi.org/10.1016/j.clsr.2022.105770

Vassos, M., Faragher, R., Nankervis, K., et al. (2024). The ethical, legal, and social implications of genomics and disability: Findings from a scoping review and their human rights implications. *Advances in Neurodevelopmental Disorders, 8*, 151–166. https://doi.org/10.1007/s41252-023-00362-1

Veerabathiran, R., & Thomas, S. M. (2025). Disability rights and policy in Asian and African continents. In *Disability across continents: Evolving policies and cultural shifts in Asia and Africa* (pp. 27–46). Springer Nature Singapore. https://doi.org/10.1007/978-981-96-6076-6_2

Wise, P. H. (2012). Emerging technologies and their impact on disability. *The Future of Children*, 169–191. https://doi.org/10.1353/foc.2012.0002.

World Health Organization. (2020). Rehabilitation 2030: a call for action: 6–7 February 2017, Executive Boardroom, WHO Headquarters, meeting report. World Health Organization.

Chapter 8
Bioethics Futures: Inclusive Innovation and Disability-Centered Policy

Abstract This chapter explores the future of bioethics from the viewpoint of disability, examining the influence of AI, neurotechnology, and biotechnology on the possible topics of inclusion, autonomy, and justice. It claims that the classic model of bioethics has always interpreted disability as a deficiency, hence, marginalized disabled people. This chapter introduces the paradigm of inclusive innovation that represents the intervention of using disability rights, the lived experience of people with disabilities, and social justice as a tool for inclusion, the intervention of ableist practices in the area of technology, data governance, and policy decisions through the promotion of justice for people with disabilities. The primary ethical issues related to algorithmic bias, mental privacy, neurotechnology, and the exclusion of people with disabilities from research and decision-making processes are discussed. The chapter also presents the disability-centered bioethics principles, in which participatory research, epistemic justice, and structural change are the central values. It finally provides the guidance for policy-making that helps to make the technology futures not only fair and accessible but also rights-based. This chapter urges the rethinking of bioethics as a tool for technology governance that is inclusive, just, and disability-informed.

Keywords Disability-centered bioethics · AI ethics · Neurotechnology · Algorithmic bias · Mental privacy · Epistemic justice

A. Shibi Anilkumar, R. Veerabathiran, *Bioethics and Disability*, SpringerBriefs in Modern Perspectives on Disability Research,
https://doi.org/10.1007/978-981-95-8197-9_8

8.1 Introduction

Artificial Intelligence (AI), neurotechnology, and biotechnology are rapidly developing technologies that are dramatically changing healthcare, research, and even disability interventions worldwide. With the help of these innovations, patient treatment and care have been more personalized through various methods, and at the same time, the boundaries for people with disabilities have been contested and pushed forward (Alanazi et al., 2025; Kargbo, 2024; Clapton, 2014).

The conventional bioethics models have generally been very much negligent of the opinions and viewpoints of disabled people, thus making it difficult to grasp the concept of disability mainly as a medical or pathological condition instead of recognizing it as a human diversity aspect. These kinds of frameworks typically indicate the deficits and vulnerability, thus overlooking the social, political, and environmental factors that determine the life of disabled people and their quality of life. Disability ethics, nonetheless, points out the flaws in traditional societies by demanding and putting into practice disability rights, personal experiences, and social justice as the main topics of moral discussion. This radically different approach to bioethics not only rejects all forms of paternalism but also promotes autonomy, agency, and equality in the process, thus facilitating the development of inclusive healthcare and policies that oppose systemic barriers and discrimination (Aarons, 2020; Vanaken, 2022; Francis & Silvers, 2016).

Inclusive innovation has a major contribution to this transformation as it allows the tech development and all its benefits to be available and affordable to the less privileged groups in society, thus creating conditions for equal growth. Moreover, it encourages coming up with and using green technologies that are not only environmentally friendly but also have economic benefits for all classes (Hardi et al., 2025). Detractors of traditional innovation patterns, particularly the unfair sharing of STRI among the various stakeholder groups, resulted in the rise of inclusive innovation as one of the new approaches. Such measures lure in active community involvement, and therefore, the poor can directly participate in the whole process of the technology's planning and execution (Morales et al., 2025).

The landscape is argued to be such that disability-related technological innovation is still absolutely crucial to the complete integration of the disabled into society. When new technological solutions emerge within a paradigm of social model, which recognizes the social and attitudinal barriers, they can be a big step forward in disability rights and participation (Breznitz & Zehavi, 2024). However, research and design aimed at the disabled community, despite having good intentions, still produce unintended harm because of the continued pervasive nature of ableism in the technological, cultural, and social systems (Shew, 2020).

The health innovation landscape is undergoing a fast transformation, and the future of bioethics should be redefined to include not only the technological advancements but also the concepts of equity, accessibility, and justice. The employment of AI and neurotechnology gives rise to a plethora of ethical dilemmas that revolve around consent, data management, algorithmic fairness, and the risk that the design embodies, or even fortifies, the ableist outlook (Umucu, 2025; Goering et al., 2022; Zurn et al., 2022).

Thus, the disability-centered research models highlight the highest involvement, representation, and collaboration of disabled people, thus guaranteeing that their opinions are taken into account in the creation of technologies and policies that influence their lives directly. The policy frameworks at the governance level should not just conform to the global rights standards but also foresee the upcoming ethical concerns by establishing regulatory frameworks that are compassionate, transparent, and prompt in reacting to the demands of the disabled communities (Abouzaglo et al., 2025). This chapter delves into the future of bioethics, taking into account the extent to which the rights of individuals with disabilities are enforced in the context of new technologies, large-scale participatory research, and the crafting of future-proof policies that would guarantee an equitable, inclusive, and ethically sustainable technological development for the disabled community.

8.2 Artificial intelligence (AI), Neurotechnology, and Ethical Inclusion

8.2.1 Artificial Intelligence (AI) in Disability

The healthcare system is undergoing a major change due to the swift progress of new technologies in the medical field, which are not only making it more efficient but also introducing marvelous innovations already. The sector is close to a complete metamorphosis because of the top-notch technology that alters the provision of care besides elevating health of patients (Yadav, 2024; Danasekaran, 2023). Artificial intelligence (AI) is simply the ability of machines or systems to do what humans do, such as think, learn, and make decisions (Khosravi et al., 2024).

Artificial intelligence (AI) has a huge role in diagnostics and is been doing so by means of efficiency to a high degree, speedy processing and accurate predictions. Artificial intelligence (AI) instruments are able to dissect X-rays, MRIs, ultrasounds, CT scans, and DXAs, thus assisting physicians in the process of diagnosing of illnesses with speed and accuracy. Apart from imaging, AI is capable of managing patient data that is once considered astronomical and consists of 2D/3D imaging, biosignals (ECG, EEG, EMG, EHR), vital signs, demographic data, medical history, and lab results, thereby enhancing the quality of decision support and providing better predictive insights (Al-Antari, 2023). Knowledge-based decision support systems continue to be a primary driver of improvement in the productivity of practitioners and the quality of healthcare outputs in clinical decision-making (Khosravi et al., 2024).

The impact of AI on the assistive technologies (ATs) sector is very significant and very much improves the quality of life and independence for disabled and older people. Assistive technologies (Ats) are a group of tools that range from simple prostheses to top-notch mobility and communication aids and they can be used for treating physical, mental, and sensory issues (Giansanti & Pirrera, 2025). Artificial intelligence-powered devices such as lip-reading applications for individuals who cannot speak, are not only able to improve communication but also ensure better access to healthcare for all. Artificial intelligence (AI) enhances accessibility in media and entertainment through sign interpretation and audio subtitles; moreover, it aids in creating inclusive work environments through job automation and individualized adjustments to the work environment (Zallio et al., 2025).

Artificial intelligence (AI) bias remains a significant issue even today. Factors such as biased mathematical formulas, unrepresentative or incorrect training data, and biased assumptions can result in algorithmic discrimination. Even the unconscious machine-learning models may reflect the discriminatory cultural trends. Given the options made regarding the importance and value of the data, the interpretations of the data might, in fact, introduce bias, thereby keeping harmful social structures alive (Belenguer, 2022). More than a billion people with disabilities all over the world, who form one of the vulnerable groups, are the ones who suffer the most from these threats, which are imposed unjustly (Vogt, 2025).

Legal safeguards could be breached through algorithmic discrimination, which is the unfavorable treatment of individuals based on protected characteristics, such as disability. The AI Bill of Rights emphasizes proactive steps to mitigate discrimination and ensure the equitable design of the system (Foley & Melese, 2025). Nevertheless, the power hierarchies inherent in AI systems remain unchallenged, even with the most diverse and extensive training datasets. Kamikubo et al. (2022) note that persons with disabilities continue to be neglected in data and, to a great extent, excluded from discussions about AI/ML governance. Disability justice is often limited to access-related matters, rather than broader social, political, and ethical issues (van Toorn et al., 2025).

Equity-centered AI development will, thus, be the process that will fuse together into one smooth ride along the whole trail of AI, starting from the foundation to deployment and monitoring (Osonuga et al., 2025). In addition, the application of AI-assisted technology in hospitals brings to the fore numerous concerns about patient privacy, data protection, fairness in the algorithms, and their functionality (Gkiolnta et al., 2025). Different variations of privacy risks include data misuse, breaches, and security measures that are not sufficient. In short, these challenges invite the presence of strong legal frameworks and the use of privacy-by-design approaches, instead of simply reacting to the situation once it comes up. The social and political ramifications of AI have to be acknowledged, and an approach of disability justice that focuses on the real-life experiences of disabled individuals should be the future direction. The collaboration of ethical AI practices with the law of equality is the only way to guarantee justice for the disabled communities (El Morr et al., 2024).

8.2.2 *Neurotechnology and Disability*

The combination of neurotechnology and disability brings urgent questions to discussion, like autonomy, mental integrity, and legal capacity for people with disabilities (Bariffi, 2025). The integration of AI into the human brain via neurotechnology has also made the concerns regarding ethics and law more significant. The capacity to decode brain activity raises questions of mental privacy, notably the dread of government and employer surveillance of brain signals. Witbrock and McCay (2025) contend that the existence of devices able to affect brain activity also opens the door to the manipulation of people's actions in a market or political context and the possible scenario where only the affluent will have access to cognitive enhancements, which would further divide society.

Neurotechnology deals with the exploration of tools and techniques that link the nervous system to technological elements such as computers, electrodes, and smart prosthetics. The devices may either stimulate the brain via electrical or optical methods or pick up neural impulses and convert them into instructions (Müller & Rotter, 2017). The invasive nature of neurosurgeries limits the use of such methods. Noninvasive neurostimulation techniques, however, make it possible to modulate brain activity using electrical or magnetic fields without surgery, even though most of these technologies are medical in origin. The impact of such technologies is most pronounced in rehabilitation, where progressive restoration of human capabilities is achieved through the treatment of cognitive and motor disorders (Ahmed & Muhammed, 2021).

Brain–Computer Interfaces (BCIs) stand as a cutting-edge neurotechnology for persons with extreme motor disabilities. Depending on the nature of the BCI, it could either act as a communication substitute for those who have lost their ability to speak or provide support in recovering such lost functions through rehabilitative therapy. Invasive BCIs require surgical electrode placement, whereas noninvasive systems record brain signals via the scalp. So far, the predominant research interest in BCIs has been in communication with people who have amyotrophic lateral sclerosis (ALS) (Chaudhary et al., 2016).

Some of the primary ethical concerns that arise with the influence of brain–computer interfaces are questions of autonomy, informed consent, and risk proportion, particularly in cases where surgical procedures are not medically indicated. The burden of costs and societal priorities are also factors that determine accessibility. However, it is essential to consider that end-users with disabilities should be involved in the development of neurotechnologies so that the tools are compatible with their identities and life experiences (Bockbrader et al., 2018).

Motor neuroprostheses, a category of neuroprosthetics, aim to restore movement via electrical stimulation of nerves, the spinal cord, muscles, or brain circuits after a stroke or spinal cord injury (Gupta et al., 2023). The rapid pace of translational research includes clinical trials investigating the possibility of direct communication between the nervous system and human–machine systems to restore lost functions (Ienca & Andorno, 2017). Functional electrical stimulation (FES), a neurostimulation method approved for clinical use, engages the motor nerves to assist or rehabilitate movement. Functional electrical stimulation (FES) can be applied via noninvasive, percutaneous, or implanted electrodes and may lead to the formation of new brain connections, depending on the level of impairment (Gupta et al., 2023).

Neuroprosthetic advancements for sensory impairments include cochlear implants (CIs) and retinal implants. Cochlear implantation has been an excellent support for the rehabilitation of children with hearing impairment, including those with the additional disability of visual impairment, as seen in Usher syndrome. Early diagnosis is essential since additional disabilities may complicate the assessment process (Eze et al., 2013). Experimental retinal implants create "phosphenes," which provide artificial visual perception for people who are blind if their optic nerve remains healthy. Users' experiences range from quick, positive adaptation to initial discomfort; however, low trust is uncommon (van Stuijvenberg et al., 2024).

Neuroscience technologies are not only providing clinical care but also augmenting human cognitive capabilities more and more. Neuroergonomics applies to the field of human performance enhancement in everyday and work situations, and BCIs are also being used to assist nondisabled individuals in decision-making, making it harder to draw a line between rehabilitation and enhancement (Cinel et al., 2019).

The technology has great potential, but it has also raised many doubts. A possible future European ethics report, published in 2005, noted that the long-term health effects of ICT implants were unclear and that this remains a gap today. Implants can worsen or cause biological rejection (Whittaker, 2005). The risks are linked to the intense ethical issues of personhood, personal identity, autonomy, and dignity. Neurotechnological interventions should be carried out with informed consent and should not violate autonomy or responsibility; interventions that threaten personhood are ethically impermissible (Müller & Rotter, 2017).

From the perspective of human rights, collecting data from the brain infringes on the fundamental rights of freedom of conscience and thought that international law provides. Since the brain's data can reveal the most intimate parts of a person's psyche, mental privacy differs from conventional privacy. Hence, more potent legal protection is called for in the neurotechnology age to ensure autonomy, dignity, and mental integrity (Szoszkiewicz & Yuste, 2025).

8.2.3 Ethical Inclusion in the Age of AI and Neurotechnology

Artificial intelligence (AI) and neurotechnology share a variety of ethical concerns, such as transparency, fairness, and value alignment, but neurotechnology poses more severe dangers to mental privacy, cognitive liberty, and identity. Neurotechnology is regarded as very important due to the fact that it can control the human brain with great precision so the individual's thoughts, emotions, and even choices will be influenced. Such powers will likely bring bioethics debate deeper, where issues like autonomy, personhood, and consent will be discussed, especially for disabled people who are already subjected to structural inequalities and discrimination (van Stuijvenberg et al., 2024).

Through the lens of bioethics that prioritizes the rights of people with disabilities, the call for ethical inclusion necessarily acknowledges the degree to which individuals with neurological diversities or hidden disabilities might find themselves at the periphery owing to the standard beliefs that are inherent in the data, algorithms, and interfaces of neurotechnology. Issues like imbalanced datasets, inadequately tailored device design, inequitable distribution, and cultural bias may lead to the oppression of these groups if they are not actively battled against (De Micco et al., 2024). Moreover, neurotech has the potential to pose privacy risks to

Table 8.1 Summary of the benefits, risks, and ethical priorities associated with artificial intelligence (AI), neurotechnology, and disability-centered inclusion

Domain	Benefits	Key risks	Ethical priorities
Artificial Intelligence (AI)	Faster and more accurate diagnosis (imaging, biosignals, EHR).	Algorithmic bias and exclusion of disabled people from data.	Inclusive datasets; bias audits.
	Strong decision-support tools.	Privacy breaches and data misuse.	Privacy-by-design.
	Enhances assistive technologies (communication aids, accessibility tools).	Reinforcement of existing inequalities.	Disability representation in AI governance.
Neurotechnology	BCIs for communication and mobility.	Threats to autonomy, mental privacy, and identity.	Protection of neurodata and cognitive liberty.
	Neuroprostheses and FES for motor/sensory restoration.	Surgical and long-term health risks.	Strong informed consent.
	Cochlear/retinal implants are improving sensory access.	Unequal access due to cost and complexity.	Co-design with disabled users.
Shared AI + Neurotech Concerns	Improve independence, rehabilitation, communication, and inclusion.	Limited transparency; embedded cultural bias.	Multistakeholder governance.
		Unequal distribution of benefits.	Fair, accessible, and culturally sensitive design.
Disability--Centered Inclusion	Ensures technology fits the real needs and experiences of disabled users.	Ableist assumptions in data, design, or policy can exclude or harm.	Apply UNCRPD principles, Prioritize autonomy, dignity, agency, and usability.

vulnerable groups, as neurodata can be identified, is challenging to de-identify, and is sometimes collected unconsciously or without meaningful consent (Table 8.1).

The ethical paradigms of AI, such as bias reduction, participatory design, transparency measures, and governance systems, can offer little to no help in neurotechnology scenarios. Neurotech has more complex needs, such as robust protections for mental integrity, identity, and agency, along with socially sensitive fairness standards and higher levels of multi-stakeholder participation. Technological advancements can be pursued only through the involvement of a wide range of people, including those with disabilities, medical and healthcare professionals, ethicists, neuroscientists, and historically excluded communities. The ethical inclusion that puts to the test autonomy, equity, and the core experience of the users in design, deployment, and governance not only protects the rights of disabled persons but also makes them stronger through the development of AI and neurotechnologies (van Stuijvenberg et al., 2024).

8.3 Disability-Centered Bioethics Research

The foundation of disability-centered bioethics is the recognition that disabled individuals have the most knowledge about their condition and, as such, should lead discussions on the ethics of health, science, and technology (Guidry-Grimes, 2022; Garland-Thomson, 2012). Through this approach, the notion that disability is a medical flaw which treatment can eliminate is completely thrown overboard; rather, disability is understood as a matter of the interplay among the physical, psychological, and social factors, and the distinction of the disabled people as beings of worth with the ability to reach their potentials has thus been made clear (Garland-Thomson, 2012; Shakespeare, 2013). The transformation, however, did not stop there; it also moved the principal questions of bioethics from "What to do to avoid or eliminate disability?" to "What adjustments are necessary in terms of systems, technologies, and policies to allow disabled people to live their lives according to their preferences, with full respect for their human dignity, and in connection with others?" (Ouellette, 2011).

Disabled individuals have been the subject of traditional bioethics discussions most of the time, and these discussions have been illustrated mainly through medical, genetic, or philosophical lenses that view disability as pain, burden, or risk (Guidry-Grimes, 2022; Kuczewski, 2001). The disability-centered approach reverses the trend and argues that "first-person" perspectives of disabled people, their political battles, and their creative coping techniques must be taken into account in ethical analysis (Kafer, 2013; Wendell, 2013). It undermines the belief that "good life" automatically has to be healthy and being the same as others in physical or mental aspects, and looks at disability as one of the instances of human diversity that can coexist with happiness, wellness, and giving (Albrecht & Devlieger, 1999; Garland-Thomson, 2012).

Research that is centered on disability also questions the social and historical forces that constitute the understanding of disability. It further analyzes the impact of factors such as colonialism, racism, sexism, and class inequality on the entire process of labeling, confining, or isolating, and on giving priority to different groups in healthcare or technology (Degener, 2016; Sabatello & Schulze, 2013). Thus, for instance, Indigenous, racialized, and rural disabled people are affected not only by the generic disability classification but by several overlapping layers of barriers to access services, participation in research, and policy making that are most often the least heeded (Aarons, 2020; WHO & World Bank, 2011). Consequently, disability-centered bioethics does not consider disability as a neutral biomedical label but rather as a social and political status that is inherently intertwined with power (Reynolds, 2018; Shakespeare, 2013).

This shift in viewpoint repositions the main ethical principles, such as autonomy, vulnerability, and justice. Autonomy is no longer a characteristic of a solitary human being making logical decisions alone; it is now a social and dependent phenomenon influenced by access to communication, support, and empowering environments (Mackenzie, 2014; Guidry-Grimes, 2022). Vulnerability is not linked to persons with disabilities exclusively, but its existence is acknowledged to be a result of or a situation where it is aggravated by the existence of inaccessible infrastructure, prejudiced attitudes, or adversarial policies (Scully, 2014; Stramondo, 2016). Justice is no longer considered to require only minor adjustments to otherwise exclusionary systems, but rather is seen as a demand for radical restructuring—redistribution of resources, rethinking of the built and digital environments, and changing institutional cultures (NCD, 2019a; UN, 2006).

Research on disability from a bioethical perspective has been conducted by placing participatory, co-produced knowledge practices at the center. Not only are disabled individuals the subjects of the research, but also the investigators, the authors, and those deciding the topics to be researched (Barnes & Mercer, 2010; Oliver, 1992). Participatory action research, community-based research, and disability-led advisory boards become standard, not exceptional, altering which questions get asked and how evidence is interpreted (Reynolds & Wieseler, 2022; Moors, 2023). The experience of life has been described as an essential kind of expertise that is in collaboration with, rather than in competition with, clinical and scientific knowledge (Nelson, 2020; Ouellette, 2011).

Moreover, such studies question everyday places of moral decision-making, which are sometimes neglected by standard bioethics. From clinical triage policies, newborn screening programs, genetic counseling practices, and algorithmic risk scores to assistive technologies and long-term care regimes, all these become the main areas of research in the disability-centered approach (HHS/OCR & ACL, 2025; Howell, 2021). One of the questions the researchers put forward is whether implicit ableism influences these practices, for instance, through very positive or very negative assumptions about "quality of life," along a continuum to attitudes of others expecting and demanding independence and productivity (Reynolds, 2018; Lagu, 2021). Besides, they examine whether the use of policy instruments such as cost-effectiveness metrics or quality-adjusted life year (QALY) models may unintentionally lead to the devaluation of disabled lives and the rationalization of governments' underinvestment in accessibility and support (Cookson et al., 2017; NCD, 2019b).

The use of a disability-centered approach is essential to the future of many technical developments, such as artificial intelligence (AI), robotics, neurotechnology, and precision medicine. The companies that produce these innovations often describe them as the next step toward overcoming the disease or even as a way to consider the person disabled. Nevertheless, the technologies can also produce the same consequences as the old standards of normality, narrowing them, increasing control over people, or even causing dependency and exclusion in new ways (De Micco et al., 2024; WHO, 2021). Disability-centered bioethics raises questions about the extent to which such technologies are conceived with disabled individuals in mind, whether the process is characterized by paternalism, and whether the technologies are made available to and affordable for diverse socioeconomic and cultural contexts (Garland-Thomson, 2012; Stahl et al., 2016). Additionally, the field of bioethics is pointing out the irregular but positive ways in which disabled persons are adopting such technologies not as the ruling but in the most creative way possible—for instance, through the digital channels of collective care, mutual aid, and political organizing (Kafer, 2013; Sabatello & Schulze, 2013).

Disability-centered bioethics has, among its priorities, the concern of epistemic justice. It questions who is believed, whose narratives are accepted as proof, and whose interpretations help determine policy. In the past, the accounts of people with disabilities regarding their abilities, wants, and living conditions were often neglected or communicated through nondisabled professionals (Asch, 2001; Wendell, 2013). Disability-centered studies purposely rectify this unequal situation by creating the analytic worldviews and institutional frameworks that will empower the disabled group, including within ethics committees, review boards, and policy forums (UCLA Disability and Health Ethics Lab, 2023). This means re-evaluating the ethics approval mechanism, providing training to ethics committees on how to spot ableism, and modifying the way data is collected so that not only the easy-to-quantify aspects are taken into account but also what is of significance to the disabled communities (NCD, 2019a; WHO & World Bank, 2011).

This method for the future of bioethics is not a niche specialization but rather an essential reorientation. It dares bioethics to consider disability as a litmus test for the wider ethical commitments to equality, solidarity, and pluralism (Guidry-Grimes, 2022; Stramondo, 2016). If the ethical frameworks are unable to include disabled people as full-blown, rights-possessing individuals, who have complex identities and dreams, then such frameworks are incomplete not only for disabled people but also for the entire society (Degener, 2016; UN, 2006). Thus, disability-centered bioethics research performs the dual role of critique and constructive imagination, a path leading to the fantastic future in which medicine, science, and technology are accountable to the people they affect and are designed to maintain the diversity of human life (Garland-Thomson, 2012; Kafer, 2013).

8.4 Policy Recommendations and Future Directions

Policies based on disability-centered bioethics should not only require compliance but also transform health, research, and innovation systems with the support of the disabled community (NCD, 2019a; UN, 2006). Regulations and policies should consider persons with disabilities not only as care recipients but also as rights holders, co-governors of technology, and designers of future health systems (Degener, 2016; Sabatello & Schulze, 2013). This will require the presence of disability expertise at all policy-making levels—from agenda setting and consultation through drafting, implementation, to evaluation—and it must be guaranteed that the disabled community is well-equipped and has the power to influence the decision rather than just giving their advice (HHS/OCR & ACL, 2025; WHO & World Bank, 2011).

A top priority is to ensure that disability rights and anti-ableism are integrated into the main bioethical and health governance frameworks. In this respect, national bioethics councils, research ethics committees, and regulatory agencies should open their doors to several disabled people and support their participation through various means, including the provision of accessible materials, sign language interpretation, and compensation for time and expertise (Reynolds, 2018; Howell, 2021). It should be made crystal clear in the ethical guidelines that discrimination on the grounds of disability is an issue in clinical decision-making, organ transplantation, access to life-sustaining treatments, reproductive technologies, and end-of-life care (Asch, 2001; Howell, 2021). It should be a rule that decisions about "futility" or quality of life can never be based on the false notion that disabled lives are less valuable or less worth preserving (Guidry-Grimes, 2022; Mukherjee et al., 2022).

Equally important is the redesign of health services and research systems to ensure that all people, regardless of their status or ability, receive the same quality of care and are allowed to participate. Aiming for universal access, strong measures such as those outlined by the UN (2006) and the WHO and World Bank (2011) should be taken across physical environments, information and communication technologies, diagnostics, and emergency plans. In addition, mandatory courses on disability, intersectionality, and anti-ableism should be offered to everyone working in healthcare systems, including doctors, nurses, and those involved in ethics and legal matters. Courses should be developed in full cooperation with people with disabilities and relevant advocacy organizations (Aarons, 2020; Reynolds, 2018). In addition, funding strategies should reward healthcare facilities and universities that are actively pursuing programs to increase disability access and participation with grants and other financial incentives (HHS/OCR & ACL, 2025; NCD, 2019a).

Policy-making regarding innovation and the not-too-distant future of technology needs to include clear disability-centric goals and safeguards as a bottom line. The law for AI, digital health, robotics, neurotechnology, and genetic interventions should require disability impact studies and ongoing community engagement (De Micco et al., 2024; WHO, 2021). For instance, such studies might question if the algorithms are indeed making biased assumptions against people with disabilities, if the assistive robots have instead caused the humans' evictions than adding to their learning process in terms of independence, or if the genetic screening programs are applying mild pressure on future parents to adhere to a very narrow idea of the "normal" (Ouellette, 2011; Stahl et al., 2016). Funding for innovation from the public treasury should give the first place to the projects that are developed in close collaboration with the disabled communities, that emphasize the aspects of access and interdependence instead of the "cure at any cost" mentality and that would be disseminated in a manner that is accessible to the most marginalized groups, including the Indigenous, rural, and low-income populations with disabilities (Sabatello & Schulze, 2013; WHO & World Bank, 2011).

Fourth, the orientation of data and evaluation systems should be changed to measure what disabled persons consider essential. Typical performance indicators like efficiency, throughput, or crude mortality are sometimes oblivious to accessibility, autonomy, and inclusion (Cookson et al., 2017). The government should mandate the creation of indicators to monitor accessible communication, shared decision-making, continuity of support, and the degree to which people live in the community of their choice, which are co-developed with disability communities and used to hold institutions accountable (UN, 2006; WHO & World Bank, 2011). The data protection and privacy regimes also need to ensure that disability-related information will not be misused for surveillance, exclusion, or discriminatory insurance and employment practices (Sabatello & Schulze, 2013; WHO, 2021).

Fifth, international and cross-cultural views on disability and bioethics must serve as guiding principles for the governance of the world and its regions. It is not reasonable for rich countries to impose their particularities regarding disabled people or the technological support to them as a common practice for all (Aarons, 2020; Degener, 2016). Instead, the global health and ethics initiatives should support disability-led, located, and rooted processes shaped by indigenous knowledge systems, community care practices, and respective cultural conceptions of the body, mind, and interdependence (Sabatello & Schulze, 2013; WHO & World Bank, 2011). International legal instruments can be instrumental in setting the lowest non-discrimination and participation standards while, at the same time, promoting diverse care, support, and innovation models that respect different human modes of existence (Garland-Thomson, 2012; UN, 2006).

In the future, creating "bioethics futures" that are genuinely inclusive will necessitate the input of disabled leaders, knowledge, and institutions. This means financing disability studies and disability bioethics courses, enabling disabled researchers and policymakers through mentorship, and setting up permanent disability advisory councils within health and science ministries (Reynolds & Wieseler, 2022; UCLA Disability and Health Ethics Lab, 2023). Additionally, it will require a shift in public perspectives on disability in education, the media, and political debates—from sorrow and liability to variety, artistic expression, and shared obligation (Kafer, 2013; Shakespeare, 2013).

Disability-centered policy, in the end, is not just about correcting the injustices of the past; it is about dreaming and building societies where the health systems, the technologies, and the moral frameworks are all based on the idea that people with disabilities are part of the society (Guidry-Grimes, 2022; Mukherjee et al., 2022). When the policy takes the disability issue as the primary dimension of justice and innovation, it can initiate the changes that will, in the long run, also benefit many other groups: the elderly, persons with chronic diseases, caregivers, and anyone whose life at a particular stage deviates from the idealized criteria of independence and productivity (UN, 2006; WHO & World Bank, 2011). Thus, disability-centered bioethics and policy are not merely a concern of a particular group, but they are the most important pathways for the eventual realization of more just, resilient, and humane societies (Garland-Thomson, 2012; NCD, 2019a).

8.5 Conclusion

The future of bioethics is expected to change the landscape of disability, abolishing traditional, medically driven interpretations and putting human rights, participation, and justice at the base of the new frameworks. The proper use of AI and neurotechnology can bring about a revolution in the way people with disabilities cope with ailments, and the issue of inequality might even be aggravated if those technologies are not developed with disabled people's viewpoints taken into consideration. The bioethics of disability changes the notions of autonomy, vulnerability, and justice by integrating the life experience and co-creation process into ethical decision-making. Inclusive innovation requires the restructuring of healthcare systems, data management, research methodologies, and the regulation of technology. The policies must ensure representation, availability, and protection against discrimination while allowing the disabled to be co-rulers of the technical future. Through the inclusion of disability in the bioethics discourse, the world will not only create more ethical systems but also more fair, strong, and empowering ones. Ultimately, disability-centered policy is the gateway to the future of ethical and inclusive technologies.

References

Aarons, D. E. (2020). The disability-rights perspective within the bioethics agenda. *Nursing Ethics, 27*(4), 1056–1065. https://doi.org/10.1177/0969733020906599

Abouzaglo, S., Patel, D., & Annaswamy, T. (2025). Inclusivity in Design: Integrating human-centered methodologies in disability research. *Disability and Health Journal, 18*(4), 101840. https://doi.org/10.1016/j.dhjo.2025.101840

Ahmed, A. A., & Muhammed, R. A. (2021). Accessibility, use and effectiveness of neurotechnology devices for improved productivity in workplace. *International Journal of Scientific and Research Publications, 11*, 15–22. https://doi.org/10.29322/IJSRP.11.10.2021.p11803

Alanazi, A. S., Salah Alanazi, A., & Benlaria, H. (2025). Enhancing healthcare for people with disabilities through artificial intelligence: Evidence from Saudi Arabia. *Healthcare (Basel, Switzerland), 13*(13), 1616. https://doi.org/10.3390/healthcare13131616

Al-Antari, M. A. (2023). Artificial intelligence for medical diagnostics-existing and future AI technology! *Diagnostics (Basel, Switzerland), 13*(4), 688. https://doi.org/10.3390/diagnostics13040688

Albrecht, G. L., & Devlieger, P. J. (1999). The disability paradox: High quality of life against all odds. *Social Science and Medicine, 48*(8), 977–988.

Asch, A. (2001). Disability, bioethics and human rights. *Handbook of disability studies, 307.*

Bariffi, F. (2025). *Mind, machine, and the law: Reimagining neurotechnology governance through disability rights.* Available at SSRN 5240404. doi:https://doi.org/10.2139/ssrn.5240404.

Barnes, C., & Mercer, G. (2010). *Exploring disability* (2nd ed.). Polity.

Belenguer, L. (2022). AI bias: Exploring discriminatory algorithmic decision-making models and the application of possible machine-centric solutions adapted from the pharmaceutical industry. *AI and Ethics, 2*(4), 771–787. https://doi.org/10.1007/s43681-022-00138-8

Bockbrader, M. A., Francisco, G., Lee, R., Olson, J., Solinsky, R., & Boninger, M. L. (2018). Brain computer interfaces in rehabilitation medicine. *PM&R, 10*(9), S233–S243. https://doi.org/10.1016/j.pmrj.2018.05.028

Breznitz, D., & Zehavi, A. (2024). Promoting inclusive innovation for disabled people in four countries: Who does what and why? *Disability and Society, 39*(4), 827–849. https://doi.org/10.1080/09687599.2022.2093698

Chaudhary, U., Birbaumer, N., & Ramos-Murguialday, A. (2016). Brain–computer interfaces for communication and rehabilitation. *Nature Reviews Neurology, 12*(9), 513–525. https://doi.org/10.1038/nrneurol.2016.113

Cinel, C., Valeriani, D., & Poli, R. (2019). Neurotechnologies for human cognitive augmentation: Current state of the art and future prospects. *Frontiers in Human Neuroscience, 13*, 13. https://doi.org/10.3389/fnhum.2019.00013

Clapton, J. (2014). Disability, ethics, and biotechnology: Where are we now? In *Voices in disability and spirituality from the land down under* (pp. 21–31). Routledge.

Cookson, R., Mirelman, A. J., Griffin, S., Asaria, M., Dawkins, B., Norheim, O. F., et al. (2017). Using cost-effectiveness analysis to address health equity concerns. *Value in Health, 20*(2), 206–212.

Danasekaran, R. (2023). The emergence of artificial intelligence in healthcare: Current trends and future directions. *Social Determinants of Health, 9*(1), 1–2. https://doi.org/10.22037/sdh.v9i1.41516

De Micco, F., Tambone, V., Frati, P., Cingolani, M., & Scendoni, R. (2024). Disability 4.0: Bioethical considerations on the use of embodied artificial intelligence. *Frontiers in Medicine, 11*, 1437280.

Degener, T. (2016). A human rights model of disability. In *Routledge handbook of disability law and human rights* (pp. 31–49). Routledge.

El Morr, C., Kundi, B., Mobeen, F., Taleghani, S., El-Lahib, Y., & Gorman, R. (2024). AI and disability: A systematic scoping review. *Health Informatics Journal, 30*(3), 14604582241285743. https://doi.org/10.1177/14604582241285743

Eze, N., Ofo, E., Jiang, D., & O'Connor, A. F. (2013). Systematic review of cochlear implantation in children with developmental disability. Otology & neurotology : official publication of the American Otological Society, *American Neurotology Society [and] European Academy of Otology and Neurotology, 34*(8), 1385–1393. https://doi.org/10.1097/MAO.0b013e3182a004b3

Foley, A., & Melese, F. (2025). Disabling AI: Power, exclusion, and disability. *British Journal of Sociology of Education, 1-22.* https://doi.org/10.1080/01425692.2025.2519482

Francis, L., & Silvers, A. (2016). Perspectives on the meaning of "disability". *AMA Journal of Ethics, 18*(10), 1025–1033. https://doi.org/10.1001/journalofethics.2016.18.10.pfor2-1610

Garland-Thomson, R. (2012). The case for conserving disability. *Journal of Bioethical Inquiry, 9*(3), 339–355.

Giansanti, D., & Pirrera, A. (2025). Integrating AI and assistive technologies in healthcare: Insights from a narrative review of reviews. *Healthcare (Basel, Switzerland), 13*(5), 556. https://doi.org/10.3390/healthcare13050556

Gkiolnta, E., Roy, D., & Fragulis, G. F. (2025). Challenges and ethical considerations in implementing assistive technologies in healthcare. *Technologies, 13*(2), 48. https://doi.org/10.1016/j.ijmedinf.2025.106051

Goering, S., Brown, T. E., McCusker, D., Montes, N., Schönau, A., Versalovic, E., & Klein, E. (2022). Integrating equity work throughout bioethics. *The American Journal of Bioethics: AJOB, 22*(1), 26–27. https://doi.org/10.1080/15265161.2021.2001108

Guidry-Grimes, L. (2022). Disability bioethics and the commitment to equality. *Theoretical Medicine and Bioethics, 43*(4), 209–220.

Gupta, A., Vardalakis, N., & Wagner, F. B. (2023). Neuroprosthetics: From sensorimotor to cognitive disorders. *Communications Biology, 6*(1), 14. https://doi.org/10.1038/s42003-022-04390-w

Hardi, I., Idroes, G. M., Márquez-Ramos, L., Noviandy, T. R., & Idroes, R. (2025). Inclusive innovation and green growth in advanced economies. *Sustainable Futures, 9*, 100540. https://doi.org/10.1016/j.sftr.2025.100540

HHS Office for Civil Rights, & Administration for Community Living. (2025). *Bioethics and access to health care for people with disabilities.* U.S. Department of Health and Human Services.

Howell, R. R. (2021). Ethical issues surrounding newborn screening. *International Journal of Neonatal Screening, 7*(1), 3.

Ienca, M., & Andorno, R. (2017). Towards new human rights in the age of neuroscience and neurotechnology. *Life Sciences, Society and Policy, 13*(1), 5. https://doi.org/10.1186/s40504-017-0050-1

Kafer, A. (2013). *Feminist, queer, crip.* Indiana University Press.

Kamikubo, R., Wang, L., Marte, C., Mahmood, A., & Kacorri, H. (2022, October). Data representativeness in accessibility datasets: A meta-analysis. In *Proceedings of the 24th international ACM SIGACCESS conference on computers and accessibility* (pp. 1–15). https://doi.org/10.1145/3517428.3544826.

Kargbo, R. B. (2024). Advancements in neurotechnology: Pioneering brain monitoring and stimulation for enhanced treatment and understanding. *ACS Medicinal Chemistry Letters, 16*(1), 20–22. https://doi.org/10.1021/acsmedchemlett.4c00584

Khosravi, M., Zare, Z., Mojtabacian, S. M., & Izadi, R. (2024). Artificial intelligence and decision-making in healthcare: A thematic analysis of a systematic review of reviews. *Health Services Research and Managerial Epidemiology, 11*, 23333928241234863. https://doi.org/10.1177/23333928241234863

Kuczewski, M. G. (2001). Disability: An agenda for bioethics. *American Journal of Bioethics, 1*(3), 36–44.

Lagu, T. (2021). *Physicians' perceptions of people with disability and their health care.*

Mackenzie, C. (Ed.). (2014). *Vulnerability: New essays in ethics and feminist philosophy.* Oxford University Press.

Moors, V. J. (2023). *Uniting disability bioethics & participatory research to ethically elucidate psychiatric conditions in persons with intellectual and developmental disabilities.* Temple University.

Morales, D., Green, L., Marlow, D., Cadena Gaitán, C., Deitrick, S., Glass, M. R., & Kempton, L. (2025). Defining inclusive innovation: Challenges and lessons for innovation districts. *Regional Studies, Regional Science, 12*(1), 740–753. https://doi.org/10.1080/21681376.2025.2543915

Mukherjee, D., Tarsney, P. S., & Kirschner, K. L. (2022). If not now, then when? Taking disability seriously in bioethics. *Hastings Center Report, 52*(3), 37–48.

Müller, O., & Rotter, S. (2017). Neurotechnology: Current developments and ethical issues. *Frontiers in Systems Neuroscience, 11*, 93. https://doi.org/10.3389/fnsys.2017.00093

National Council on Disability. (2019a). *Bioethics and disability series: Introduction.* National Council on Disability.

National Council on Disability. (2019b). *Quality-adjusted life years and the devaluation of life with a disability.* National Council on Disability.

Nelson, R. H. (2020). Now and again: Reappraising disability leave as an accommodation. *BYU Law Review, 46*, 1489.

Oliver, M. (1992). Changing the social relations of research production? *Disability, Handicap and Society, 7*(2), 101–114.

Osonuga, A., Osonuga, A. A., Fidelis, S. C., Osonuga, G. C., Juckes, J., & Olawade, D. B. (2025). Bridging the digital divide: Artificial intelligence as a catalyst for health equity in primary care settings. *International Journal of Medical Informatics*, 106051. https://doi.org/10.1016/j.ijmedinf.2025.106051.

Ouellette, A. (2011). *Bioethics and disability: Toward a disability-conscious bioethics*. Cambridge University Press.

Reynolds, J. M. (2018). Three things clinicians should know about disability. *AMA Journal of Ethics, 20*(12), 1181–1187.

Reynolds, J. M., & Wieseler, C. (Eds.). (2022). *The disability bioethics reader*. Routledge.

Sabatello, M., & Schulze, M. (Eds.). (2013). *Human rights and disability advocacy*. University of Pennsylvania Press.

Scully, J. L. (2014). Disability and vulnerability: On bodies, dependence, and power. *Vulnerability: New essays in ethics and feminist philosophy, 204*, 209–210.

Shakespeare, T. (2013). *Disability rights and wrongs revisited*. Routledge.

Shew, A. (2020). Ableism, technoableism, and future AI. *IEEE Technology and Society Magazine, 39*(1), 40–85. https://doi.org/10.1109/MTS.2020.2967492

Stahl, B. C., Timmermans, J., & Mittelstadt, B. D. (2016). The ethics of computing: A survey of the computing-oriented literature. *ACM Computing Surveys (CSUR), 48*(4), 1–38.

Stramondo, J. A. (2016). Why bioethics needs a disability moral psychology. *Hastings Center Report, 46*(3), 22–30.

Szoszkiewicz, Ł., & Yuste, R. (2025). Mental privacy: Navigating risks, rights and regulation: Advances in neuroscience challenge contemporary legal frameworks to protect mental privacy. *EMBO Reports, 26*(14), 3469. https://doi.org/10.1038/s44319-025-00505-6

UCLA Disability and Health Ethics Lab. (2023). *Disability and health ethics lab: Mission and projects*. University of California.

Umucu, E. (2025). Artificial intelligence and health equity for people with disabilities: An integrated framework for disability-inclusive AI design. *Inquiry: a journal of medical care organization, provision and financing, 62*, 469580251365472. https://doi.org/10.1177/00469580251365472

United Nations. (2006). *Convention on the rights of persons with disabilities*. United Nations.

van Stuijvenberg, O. C., Broekman, M. L., Wolff, S. E., Bredenoord, A. L., & Jongsma, K. R. (2024). Developer perspectives on the ethics of AI-driven neural implants: A qualitative study. *Scientific Reports, 14*(1), 7880. https://doi.org/10.1038/s41598-024-58535-4

van Toorn, G., Scully, J. L., & Gendera, S. (2025). "This robot is dictating her next steps in life": Disability justice and relational AI ethics. *AI and Society, 40*, 4473–4483. https://doi.org/10.1007/s00146-025-02224-x

Vanaken, G. J. (2022). Cripping vulnerability: A disability bioethics approach to the case of early autism interventions. *Tijdschrift voor genderstudies, 25*(1), 19–40. https://doi.org/10.5117/TVGN2022.1.002.VANA

Vogt, Y. (2025). Disability and algorithmic fairness in healthcare: A narrative review. *Journal of Medical Artificial Intelligence, 8*. https://doi.org/10.21037/jmai-24-415

Wendell, S. (2013). *The rejected body: Feminist philosophical reflections on disability*. Routledge.

Whittaker, P. (2005). Biotechnology and ethics / the next 10 years. In General Report on the Activities of the European Group on Ethics in Science and New Technologies to the European Commission Office for Official Publications of the European Communities. http://ec.europa.eu/european_group_ethics/publications/docs/genactrep05txt_en.pdf

Witbrock, M., & McCay, A. (2025). Don't forget the upside of neurotechnology. *AI & Society, 40*, 3375–3377. https://doi.org/10.1007/s00146-024-02161-1

World Health Organization. (2021). *Ethics and governance of artificial intelligence for health*. World Health Organization.

World Health Organization, & World Bank. (2011). *World report on disability*. World Health Organization.

Yadav, S. (2024). Transformative frontiers: A comprehensive review of emerging technologies in modern healthcare. *Cureus, 16*(3), e56538. https://doi.org/10.7759/cureus.56538

Zallio, M., Ike, C. B., & Chivăran, C. (2025). Designing artificial intelligence: Exploring inclusion, diversity, equity, accessibility, and safety in human-centric emerging technologies. *AI, 6*(7), 143. https://doi.org/10.3390/ai6070143

Zurn, P., Stramondo, J., Reynolds, J. M., & Bassett, D. S. (2022). Expanding diversity, equity, and inclusion to disability: Opportunities for biological psychiatry. *Biological Psychiatry: Cognitive Neuroscience and Neuroimaging, 7*(12), 1280–1288. https://doi.org/10.1016/j.bpsc.2022.08.008

Appendix A: Figures and Tables

1. **Figure 1.1** Historical Progression of Disability Perceptions and Treatment.
   This figure illustrates the historical evolution of disability treatment and societal attitudes toward persons with disabilities, progressing from exclusion to rights-based inclusion. The first section depicts early discrimination, neglect, and infanticide of disabled infants, symbolized by a prohibition sign over mobility aids and a distressed infant. The second section represents the emergence of religious charity and basic care institutions, as evidenced in church-based support and protective caregiving imagery. The third section highlights the rise of specialized schools, blind and deaf institutes, and asylums that, although providing structured environments, often reinforced segregation. The fourth section portrays twentieth-century injustices, including forced sterilization and deeply embedded societal stigma against disability. The final section signifies modern disability rights and bioethics, represented by the figure of Lady Justice holding scales, symbolizing equality, legal protections, and ethical responsibility. Overall, the image narrates the shift from historical marginalization to contemporary frameworks grounded in dignity, justice, and disability rights.
2. **Figure 1.2** Overview of the Medical, Social, and Bioethical Models of Disability
   The figure presents three models of disability: Medical, Social, and Bioethical, displayed as three colored panels beneath a banner titled "Different Models of Disability," with an illustration of a child above the banner. The Medical Model panel states that disability is viewed as an individual impairment requiring medical treatment, and it includes an icon of a wheelchair user with a label indicating that the individual is considered the problem. The Social Model panel describes disability as a social justice issue caused by societal barriers, shown with an icon of a person facing an obstruction. The Bioethical Model panel emphasizes equitable access, moral responsibility, and fairness in

A. Shibi Anilkumar, R. Veerabathiran, *Bioethics and Disability*, SpringerBriefs in Modern Perspectives on Disability Research,
https://doi.org/10.1007/978-981-95-8197-9

disability-related practices, accompanied by icons symbolizing support and justice.

3. **Figure 2.1.** Overview of Genetic Screening and Diagnostic Techniques.
   The figure illustrates three significant categories of genetic screening and diagnostic techniques: prenatal testing, newborn screening, and genetic testing. At the top, a stylized hand writing on a DNA helix symbolizes genetic analysis. Below this, three labelled sections contain visuals representing each testing type. The prenatal testing section includes icons of an ultrasound machine and blood sample tubes, as well as depictions of chorionic villus sampling and amniocentesis. The newborn screening section features laboratory equipment such as automated analyzers, mass spectrometers, and sample testing platforms. The genetic testing section features a DNA sequence, a marked mutation, and an illustration of an infant's internal organs, representing disease detection and molecular analysis.
4. **Figure 3.1** Layers of Bioethical Governance for Disability.
   The figure depicts three overlapping circles representing layers of bioethical governance for disability. The smallest yellow circle, labelled "Healthcare Practice," includes items such as hospital ethics committees, patient–provider decision-making, and community-based rehabilitation. This circle overlaps with a larger blue circle labelled "National Governance," which lists actions such as approving research projects, enacting disability rights laws, implementing anti-discrimination provisions, establishing accessibility standards, raising public awareness, and monitoring human rights violations. Both circles are encompassed by an even larger green circle titled "Global Ethical Frameworks," which includes institutions and guidelines such as the UNCRPD, UNESCO, WHO, the Nuremberg Code, the World Medical Association, and CIOMS international ethical guidelines. A red wheelchair-user icon appears on the left, symbolizing disability.
5. **Figure 4.1** Ethical Principles and Decision-Making Processes at the End of Life.
   The figure presents a visual overview of end-of-life (EOL) decisions, processes, and ethical principles. The top section, labelled "EOL Decisions and Processes," includes icons representing a physician consultation, shared decision-making between patient and provider, a hospitalized patient under medical supervision, a patient will document, and a comparison of medication choices with resuscitation (CPR). The lower section, titled "End-of-Life Ethics," outlines four core ethical principles. Autonomy includes the right to choose or refuse treatment, to use advance directives, and to uphold personal values. Beneficence focuses on acting in the patient's best interest and providing comfort. Nonmaleficence emphasizes avoiding unnecessary suffering and minimizing harmful interventions. Justice highlights fair access to palliative care and the absence of discrimination based on age, disability, or socioeconomic factors.
6. **Figure 5.1** Barriers to Reproductive Health for Persons with Disabilities and Their Associated Consequences.

The figure illustrates barriers to reproductive health for persons with disabilities (PwD) on the left and the consequences of these barriers on the right. The barrier section includes icons representing inaccessibility, medical bias, financial barriers, lack of institutional support, and limited access to reproductive health information or contraception. The consequences section shows icons representing unintended pregnancy, vulnerability to abuse, denial of reproductive autonomy, psychological distress, and confusion or lack of guidance. The two sections are visually divided: the barrier icons are placed under a red banner labelled "Barriers to Reproductive Health for PwD," and the consequences are grouped in a large circle under a banner labelled "Consequences."s

Appendix B: Methodology and Ethical Considerations

1. **Methodological Approaches**

   (a) Interdisciplinary Literature Review

   Chapters rely on interdisciplinary sources spanning bioethics, genetics, disability studies, clinical medicine, law, sociology, and public health, enabling comprehensive synthesis.

   (b) Case-Based Ethical Analysis

   Used in prenatal diagnosis, termination decisions, and clinical justice cases.

   c. Policy and Framework Analysis

   The chapters assess legal instruments such as the CRPD, the ADA, the GDPR, and bioethical declarations.

   (c) Comparative Legal Method

   The chapter compares disability policies across the EU, the USA, India, and South Africa.

   (d) Socio-Ethical Critique

   Used to interrogate ableism, stigma, genetic normalization, and the medical model.

2. **Ethical Considerations**

   (a) Autonomy

   Central to genetic counseling, reproductive choice, AI governance, and disability rights.

   (b) Informed Consent

   Essential in NIPT, gene editing, neurotechnology, and forensic genomics.

   (c) Justice and Nondiscrimination

   Obligations under CRPD, ADA, and global policies to prevent healthcare bias.

A. Shibi Anilkumar, R. Veerabathiran, *Bioethics and Disability*, SpringerBriefs in Modern Perspectives on Disability Research,
https://doi.org/10.1007/978-981-95-8197-9

(d) Nonmaleficence

Avoiding harms associated with embryonic selection, algorithmic bias, forced sterilization, and coercive care.

(e) Beneficence

Ensuring reproductive technologies, AI, and rehabilitation strategies improve well-being.

(f) Privacy and Data Protection

Heightened concerns in genomics, AI, and neurodata.

(g) Inclusion and Lived Experience

Emphasize participatory research and the meaningful integration of disabled communities.

Appendix C: Frameworks, Laws and Policies

1. **International Frameworks**
   (a) UNCRPD – United Nations Convention on the Rights of Persons with Disabilities
   (b) UNESCO Universal Declaration on Bioethics & Human Rights (UDBHR)
   (c) WHO Rehabilitation 2030 Framework
   (d) UNICEF Disability Inclusion Guidelines (Referenced within global disability data)
2. **Regional and National Laws**
   European Union
   (a) European Disability Strategy 2010–2020
   (b) European Disability Rights 2021–2030 Strategy
   (c) GDPR (EU Data Protection Law)

   United States
   (a) ADA (Americans with Disabilities Act)
   (b) Section 504, Rehabilitation Act (1973)
   (c) Affordable Care Act (ACA)
   (d) WIOA – Workforce Innovation and Opportunity Act

   India
   (a) Rights of Persons with Disabilities (RPwD) Act, 2016
   (b) National Trust Act, 1999

   South Africa
   (a) South African Constitution – Equality Clause (Section 9)
   (b) Promotion of Equality and Prevention of Unfair Discrimination Act

A. Shibi Anilkumar, R. Veerabathiran, *Bioethics and Disability*, SpringerBriefs in Modern Perspectives on Disability Research,
https://doi.org/10.1007/978-981-95-8197-9

3. **Bioethics and Clinical Governance**
   (a) Principle of Respect for Human Vulnerability—UNESCO IBC
   (b) Medical Futility and DNR Guidelines (Global ICU ethics standards)
   (c) AI Governance (AI Bill of Rights—U.S.)

Appendix D: Statistical Insights

1. **Prevalence and Demographics**

   Global Disability Prevalence

   (a) 16% of the world population (1.3 billion people) live with significant disabilities.
   (b) Millions of children have moderate-to-severe disabilities globally (UNICEF).

   Healthcare Inequality Indicators

   (a) PwD have higher morbidity and premature mortality rates globally.
   (b) 25–65% of disabled people in EU member states struggle to access basic services (healthcare, banking, transport).

2. **Policy and Law Implementation Outcomes**

   CRPD Adoption

   (a) 46 African countries have ratified CRPD; 33 ratified its Optional Protocol.

   Employment Gap

   (a) PwD face unemployment rates over twice those of nondisabled peers.

A. Shibi Anilkumar, R. Veerabathiran, *Bioethics and Disability*, SpringerBriefs in Modern Perspectives on Disability Research,
https://doi.org/10.1007/978-981-95-8197-9

Glossary

Disability A physical, cognitive, or sensory condition that limits participation in daily activities and social interaction.

Bioethics An interdisciplinary field examining ethical issues in medicine, biotechnology, and healthcare practices.

Eugenics A historical movement promoting controlled breeding to "improve" populations, leading to coercive sterilization and genocide.

Forced Sterilization State-sanctioned prevention of reproduction among disabled groups, historically practiced worldwide.

Institutionalization Segregation of disabled individuals in asylums or closed facilities, often involving abuse and deprivation.

Intersectionality A framework recognizing how disability interacts with gender, race, class, and other identities to create layered disadvantage.

UNCRPD (United Nations Convention on the Rights of Persons with Disabilities) A global treaty defining disability rights and requiring dignity, autonomy, and participation for PwDs.

Genetic Engineering Manipulation of DNA for therapeutic, diagnostic, or enhancement purposes.

CRISPR-Cas9 A precise gene-editing tool enabling modification of targeted DNA sequences.

NIPT (Noninvasive Prenatal Testing) A blood test analyzing fetal DNA to screen for chromosomal anomalies.

PGD (Preimplantation Genetic Diagnosis) Embryo screening during IVF to avoid hereditary disease transmission.

Precision Medicine Personalized treatment based on an individual's genetic profile.

Disability-Selective Abortion Termination based on fetal disability traits, raising ethical issues of valuation and ableism.

A. Shibi Anilkumar, R. Veerabathiran, *Bioethics and Disability*, SpringerBriefs in Modern Perspectives on Disability Research,
https://doi.org/10.1007/978-981-95-8197-9

Informed Consent A process ensuring individuals understand genetic risks, limitations, and consequences before testing.

Genetic Determinism The belief that genes alone define human traits and destinies.

Posthumanism A philosophical view questioning boundaries between humans, technology, and enhanced beings.

Transhumanism A movement supporting technological enhancement (genetic, prosthetic, cognitive).

Pharmacogenomics (PGx) Study of how genetic variation impacts medication effectiveness and tolerance.

Enhancement Ethics Debates surrounding improving human traits beyond typical functioning.

Clinical Justice Ensuring equitable access, fair treatment, and nondiscriminatory healthcare for PwDs.

Ableism Discrimination or prejudice against individuals with disabilities.

GDPR (General Data Protection Regulation) EU law protecting sensitive data, including disability and genomic information.

Inclusive Healthcare A healthcare system that eliminates barriers, biases, and inaccessible environments for PwDs.

Accessibility Designing environments, information, and services enabling full participation by all individuals.

IVF (In Vitro Fertilization) Fertilization of an egg and sperm outside the body.

ART (Assisted Reproductive Technology) Treatments manipulating gametes and embryos to achieve pregnancy.

Embryo Selection Choosing embryos based on genetic results, quality, or absence of disease.

Designer Babies A speculative concept where embryos are genetically modified for desirable traits.

Surrogacy A Process in which a gestational carrier bears a child for intended parents.

Severity Thresholds Ethical markers defining which genetic conditions justify selection or exclusion.

Soft Law Nonbinding policies that guide national disability governance (e.g., EU strategies).

Rehabilitation 2030 (WHO) A global initiative integrating rehabilitation within health systems.

Data Justice Ensuring fairness in the handling of medical and disability-related data.

Digital Divide Unequal access to technology affects the participation of people with disabilities.

AI (Artificial Intelligence) Systems capable of learning, problem-solving, and decision-making.

Brain–Computer Interface (BCI) Technology enabling brain signals to control computers or prosthetics.

Neurotechnology Tools interacting with the brain. e.g., implants, neurostimulation, decoding neural activity.

Mental Privacy A right protecting thoughts and brain data from unauthorized access.

Algorithmic Bias Systematic discrimination embedded in AI due to flawed or incomplete datasets.

The manufacturer's authorised representative in the EU is Springer Nature Customer Service Centre GmbH, Europaplatz 3, 69115 Heidelberg, Germany. If you have any concerns regarding our products, please contact ProductSafety@springernature.com

Printed and bound by CPI Group (UK) Ltd, Croydon, CR0 4YY
07/07/2026
02160931-0006